With proud thanksgiving, a mother for her children,
England mourns for her dead across the sea.
Flesh of her flesh they were, spirit of her spirit,
Fallen in the cause of the free.

*Solemn the drums thrill*; Death august and royal
Sings sorrow up into immortal spheres,
There is music in the midst of desolation
And a glory that shines upon our tears.

They went with songs to the battle, they were young,
Straight of limb, true of eye, steady and aglow.
They were staunch to the end against odds uncounted;
They fell with their faces to the foe.

They shall grow not old, as we that are left grow old:
Age shall not weary them, nor the years condemn.
At the going down of the sun and in the morning
We will remember them.

They mingle not with their laughing comrades again;
They sit no more at familiar tables of home;
They have no lot in our labour of the day-time;
They sleep beyond England's foam.

But where our desires are and our hopes profound,
Felt as a well-spring that is hidden from sight,
To the innermost heart of their own land they are known
As the stars are known to the Night;

As the stars that shall be bright when we are dust,
Moving in marches upon the heavenly plain;
As the stars that are starry in the time of our darkness,
To the end, to the end, they remain.

*For the Fallen by Robert Laurence Binyon, 1914*

# Solemn, The Drums Thrill

*Essays on the Fallen Heroes*
*of Stanwood Camano:*
*World War I to Afghanistan*

by Richard A. Hanks

Coyote Hill Press

Published by Coyote Hill Press, Camano Island, WA

Layout & Design by Robin S. Hanks


First Edition, 2020

Printed in the United States

ISBN: 978-1-7358615-1-7  All rights reserved.

Front Cover: Photograph by Barbara Joyce, Memorial Day 2020

# Dedication

The Community Veteran's Memorial built by the Stanwood Area Historical Society is dedicated to the brave sons and daughters of the greater Stanwood/Camano Island community who answered their nation's call and gave the ultimate sacrifice on distant fields. We honor all in our armed forces who leave home, hearth and family to risk their lives for the freedoms we hold dear. Memories may fade but our hearts will remember.

*"I pray that our Heavenly Father may assuage the anguish of your bereavement, and leave you only the cherished memory of the loved and lost, and the solemn pride that must be yours, to have laid so costly a sacrifice upon the altar of Freedom."*

*Abraham Lincoln to Mrs. Bixby, November 21, 1864*

# Contents

# Foreword

One morning while talking and drinking coffee with an old friend, Bill Keller, we made the decision to begin a project to construct a Veterans Memorial in Stanwood. Later, Bill approached the Stanwood Area Historical Society's historians, Dr. Richard Hanks and Bill Blandin, to discuss our concept and request assistance. After several discussions, our vision evolved: to honor all those who served, and to memorialize those from our community who died while serving during America's five major conflicts: World War One, World War Two, Korean War, Vietnam War, and the Iraq-Afghanistan Conflict. Both men graciously and enthusiastically agreed to begin the name identification research. Their painstaking efforts located the name, date of death, and limited information for forty-nine service members and one nurse from our community, who died during those conflicts.

While we focused on the planning, coordination, and construction of the Memorial, Dr. Hanks expanded their initial identification phase by meticulously researching the lives of each fallen individual. Those names, reported as combat statistics, became human beings again. The "kid" next door, or down the street, like someone we have all known, who put future aspirations on hold and enlisted in the military. Their lives changed drastically as they quickly learned structure, discipline, unity, and loyalty. Each selflessly began an arduous sacrifice and commitment during a critical time in our history. As Dr. Hanks accurately depicts, they were not characters in a movie, they were hardworking people who served their country with honor. Although the majority were enlisted soldiers, "boots on the ground", the backbone of the military, each made meaningful contributions. His essays raised awareness and helped generate necessary local interest in our project. Those fifty names are now permanently engraved on top of their respective conflict pedestal. In their honor, and for the benefit of friends, relatives, and all of us, Dr. Hanks has written a meaningful, personal description of each life, and briefly presented the circumstances that resulted in their loss.

 As a 26 year Navy veteran who flew attack helicopters in the Delta Region of South Vietnam, I am eternally grateful to Dr. Hanks for his research and his book that remembers and recognizes the sacrifice of one of my classmates and the other 49 individuals. Upon the completion of the Stanwood Area Veterans' Memorial, we acknowledged the local businesses and individuals who volunteered in so many ways to complete the project. Their efforts are proudly displayed, their names are engraved on a granite pedestal located at the Memorial entrance. But one name whose efforts were just as important, but not as noticeable, is missing on that pedestal. In truth, without the painstaking, in-depth research of Dr. Richard Hanks, the Memorial would have probably been only bricks, cement, and granite with generic comments. His tireless efforts identified the 50 names that will forever be memorialized, and they brought meaning to the sacrifices that took their lives so early. As always, he made history personal, interesting, and meaningful.

Jim Joyce

Captain, United States Navy

Retired

# Preface

The size of a community does not equal the measure of the hearts of those who live there. In times of crisis, ordinary sons and daughters of the greater Stanwood area responded with extraordinary commitment. Neither embracing the dangers of service nor shirking from them, they, often with quiet grace, accepted their moment of purpose even to the point of death. Families hold the memories of those soldiers who paid the ultimate sacrifice during conflict—their bodies lying in sacred soil stretching from the fields of France to the volcanic islands of the South Pacific, the jungles of Asia or beneath the great blue waters of the world's oceans. The idea for this book grew with the desire to pay homage to the lost and loved of the Stanwood Camano area.

The Community Veterans' Memorial built on the grounds of the Stanwood Area Historical Society was an idea four years in the making. It honors all who have proudly worn America's uniform, but it pays specific tribute to those local citizens who did not return after answering their country's call. The names of these heroes, from World War I to Afghanistan and Iraq, are etched on granite pedestals within the memorial park.

The idea of a memorial started with the passion and focus of Bill Keller and Jim Joyce who, struck by the solemn power of the Vietnam wall, realized that such a memorial to local servicemen and women of the greater Stanwood/Camano area was absent. Their appeals on behalf of the memorial brought forth the generosity of the community and we are deeply appreciative to the people and businesses of the area who donated their time, labor and resources to make this memorial a reality.

Often the sacrifice of America's veterans goes unheeded— the silent purpose of their service lost to history. As a way of generating interest in the memorial and these veterans, I began researching and writing essays on the 49 soldiers and one nurse whose names appear on the memorial's marble pedestals. Many of these accounts initially appeared over three years in the Crab Cracker magazine. My intent was to breathe life into the human stories of these fallen heroes and make their sacrifice more than names carved in stone. For those veterans from the greater Stanwood area, the memorial, and this book, marks a community's grateful appreciation.

*"Show me the manner in which a nation or a community cares for its dead and I will measure with mathematical exactness the tender sympathies of its people, their respect for the laws of the land and their loyalty to high ideals."*

*William Gladstone, 1871*

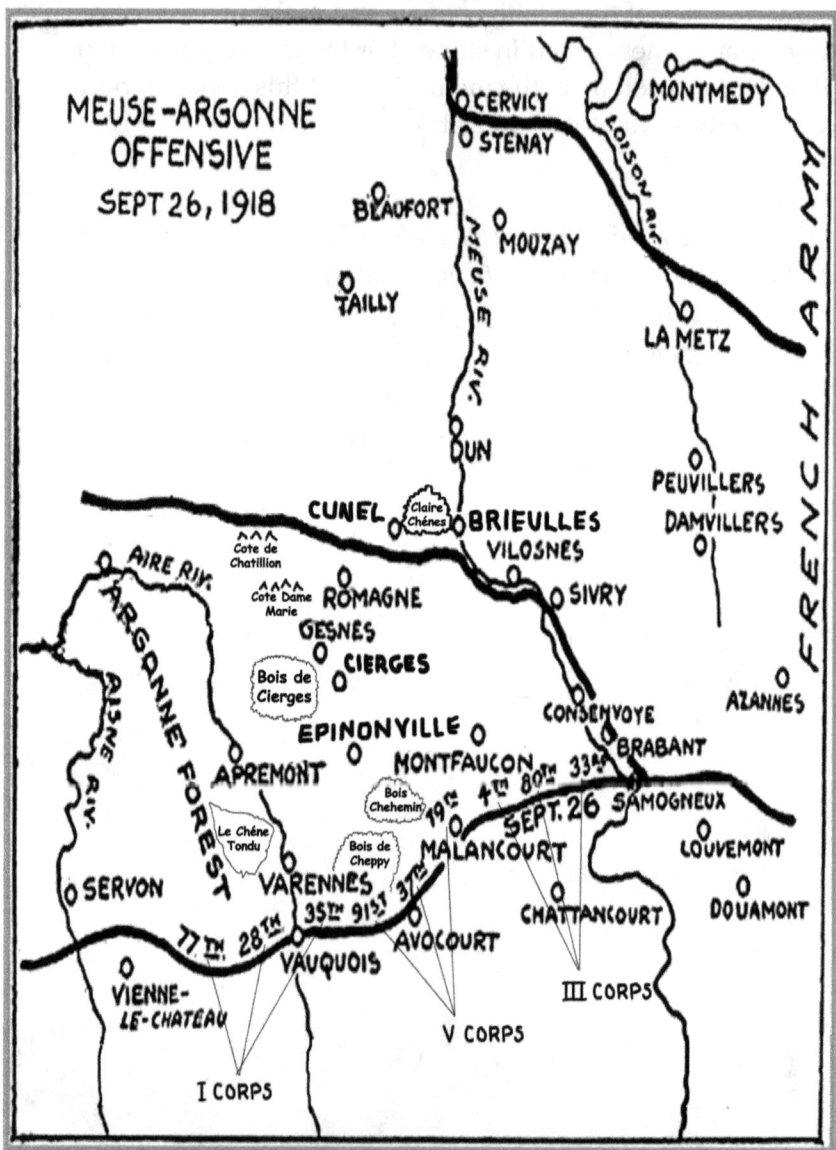

MEUSE-ARGONNE
OFFENSIVE
SEPT 26, 1918

# World War I
## The War to end all Wars

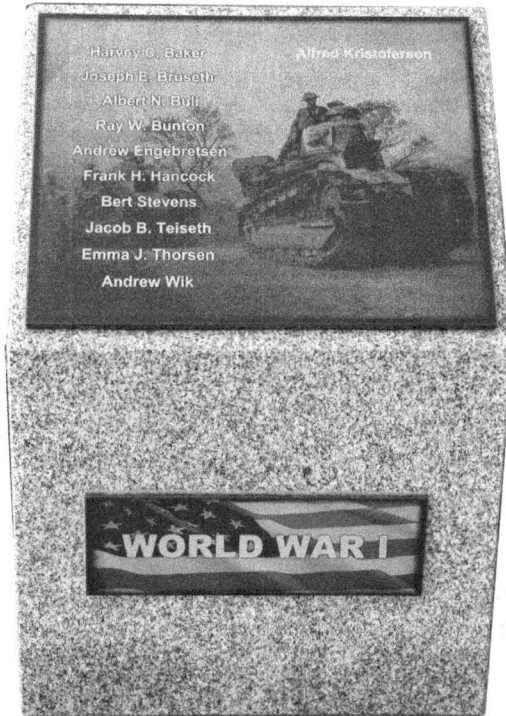

America in 1917 was on its way to urbanization but still held on to the rustic atmosphere of its founding with half of the U.S. population still living in small towns and rural areas like Stanwood, Washington. With the declaration of war on April 6, 1917 the call went out for volunteers but the appeals fell short with only 73,000 coming forward. Far less than what was needed for the war, the Congress passed the Selective Service Act of May 18, 1917 mandating registration by American males 21 to 30 years of age. The vast majority of men complied and registered for the draft. In the end, 70% of those in the U.S. Army would be conscripts, roughly 2.8 million men. One week after the declaration of war, the Committee on Public Information was created to convince Americans to fully support and participate in the war effort.

America had pulled back from the more expansionist policies of Theodore Roosevelt and William Howard Taft following the Spanish American War in 1898. Woodrow Wilson sought to influence the world more by moral example rather than pressing American interests by economic or military pressure. America's policy was one of neutrality. This would prove more difficult as the effects of war, such as the sinking of the Lusitania in 1915, came ever closer.

Americans were still very much a people of place—the one they could see outside their windows. In the Pacific Northwest, many had come from northern Europe and Scandinavia specifically. World War I would shake the nation from its roots as a home of individual promise and progress to one of awareness of the outside world, its trouble and the collective demands imposed. The country still had a foot in the 19th century even while awakening to the 20th.

All but one of the young men in Stanwood answered quickly the call of the country—many still juniors and seniors in high school. The one who resisted, was called out in the local paper but not identified, a minor omission in a small town like Stanwood because "everybody knows who he is," the newspaper proclaimed, "and so, to leave this page unsullied his name will not be printed. His yellow streak speaks for itself" (*Stanwood Tidings*, April 20, 1917). The Red Cross gave a farewell party for the departing recruits in the spring of 1917 at the Odd Fellows Hall in Stanwood which was decorated with American flags (today's Floyd Norgaard Cultural Center). On the day they left for training, a local band played as forty carloads of relatives and well-wishers gathered at the train station. Of the ten men and one woman from the greater Stanwood area who responded over the next several months, some willingly and others less enthusiastically, all but two would die in the war's final great struggle, the Meuse Argonne offensive (September 26—November

11, 1918). The battle had three distinct phases. Stanwood area citizens would make sacrifices in all three. Three of the ten would fall in a different battle; one as lethal even if not seen.

History's greatest pandemic, dubbed the Spanish Flu, raged across continents starting in the spring of 1918. It spread quickly in military camps and packed troop ships in particular. However, the origin of the disease was not Spain but rather historians cite the first case in the United States occurring at Camp Funston, Kansas in the spring of 1918. A nurse from Silvana serving in nearby Camp Dodge would fall victim just days before the end of the war. Spain's name was attached arbitrarily to the virus only because its newspapers were the first to report on the scourge in May 1918. As a neutral country in the war, Spain's media was not under the same rules of censorship as warring nations which suppressed news of the disease so as not to damage public morale.

# Called to the Colors
## Harvey C. Baker

The world of Harvey Charles Baker changed drastically in 1917. At the time the 23-year-old Baker was working at Esary Camp #2 on Camano Island—today's Cama Beach. The camp owned by the Esary brothers, James and Thomas—another brother Andrew died in a logging accident in 1904--operated a lumber mill in the area and Harvey worked cutting shingle bolts. Harvey was among the first taken in the draft for World War I and was the first from Island County "called to the colors" to die. Not from a bullet fired in anger but from a disease methodically stalking every military camp.

The young man, born in Duluth, Minnesota in 1892, spent much of his formative years in the territories of Alberta, Canada where his father, a farmer, moved the family. By 1911 father and son had both found employment in the area's coal fields. It was the same year Harvey's mother died. By the time of his registration, he was working the logging camps of Island County.

Harvey'a request for an exemption as the sole supporter of his brother and sister, failed to exclude him from the draft. He was sent to Fort Lewis and assigned to the 161st Infantry Regiment, part of the 41st ("Sunset") Division which shipped to France in November 1917. The division was first organized on 18 September 1917 as the 41st Division at Camp Greene, North Carolina. Under the command of Major General Hunter Liggett, the division was largely made up of National Guardsmen from the northwestern U.S., including Oregon, Washington, Idaho, and Montana. The division included the 161st, 162nd, 163rd, and 164th Infantry Regiments. Once in France, the Division was designated the 1st Depot Division used to provide replacements to other units on the front lines. Harvey would not live long enough to see the front lines, however.

Infectious diseases such as measles, mumps and tuberculosis and later influenza compromised the health of thousands in America's training camps. By war's end four million men would be hospitalized—86 percent of those due to disease. Most American soldiers died not on battlefields but in hospitals during World War I despite the advancement in medicine mainly due to the pandemic of influenza—over 63,000 out a total of 116,516 men who were lost.

 Disease spread easily in the crowded conditions aboard the transports for the 41st Division, with a number of soldiers developing cases of pneumonia. After landing, conditions were bleak in the Division's camp. The weather was bitterly cold and supplies were scarce as were proper sanitation facilities. Many of the sick were sent to an army hospital in La Courtine. It is likely that Harvey Baker was with them. The hospital consisted of an empty barrack. Most medical supplies had been diverted to other units once ashore at the port of Brest. On January 24, 1918, the division was sent to St. Aignan-Noyers where a report noted the immediate beneficial effect upon the health of the command. Too late for Private Baker, however, who succumbed to pneumonia twenty days before the move. No glorious action under fire took Harvey Baker's life. His loss was a lonely sacrifice to the ravages of disease in a foreign land. He was buried in the American cemetery in St. Mihiel, Lorraine, France after his death in January 4, 1918 and lies there still.


# A Village Too Far

## Andrew Wik


Andrew Wik died in the first phase of the Meuse Argonne offensive. That phase beginning September 26, 1918 until October 1st, also took the lives of Frank Hancock and Bert Stevens. It's unknown when Andrew Wik came to the U.S. but he settled in the area of East Stanwood. His cousin Andrew Harnvik joined him there after arriving in New York City in April 1909 from their mutual hometown of Aure, Norway. Andrew Wik may have changed his name since records list Peder Harnvik as his father.

Andrew Wik found employment working the rich waters off Alaska. He registered for the draft in July 1917, citing East Stanwood as his residence, but at the time was working for the Sockeye Salmon Company's new cannery plant along Morzhovoi Bay in the Aleutian Islands. It is likely that Andrew was a crew member of the steamer *Prince John* which carried cargo and passengers from Prince Rupert, British Columbia to Skagway, Alaska a year earlier.

Andrew Wik was still a Norwegian citizen when he was inducted into the U.S. Army. He reported to Camp Lewis, Washington and was assigned to Company F, 362nd Infantry Regiment, 91st Infantry Division—the "Wild West Division." Most of the 91st consisted of men from the Northwest. In June 1918, the Division boarded trains for Camp Merritt, New Jersey to be outfitted and prepared for France. The first units left July 6, 1918. Three days before this departure, Private Andrew Wik became a naturalized U.S. citizen.

Wik and the 362nd trained for six weeks upon arriving in Montigny Le Roi, France. They became a reserve regiment at the battle of St. Mihiel, an American victory, during the second week of September. Moving to the area of Meuse-Argonne after September 13, they were placed, for the first time, along the front lines; the 91st division was deployed at the center of the American positions—the 362nd on the right flank of the division. The area was a bleak, muddy landscape of shell craters inundated by intermittent torrents of rain. The threat of nightly artillery gas attacks kept the men on edge, often attempting restlessly to sleep in their protective masks. A 1920 regimental history remembered: "Certain it is that the sleep we had was broken many times the first few nights by cries of 'Gas' from thousands of throats..."

In the night before the attack on September 26, the heavy salvos of American artillery made rest impossible. Men smoked, wrote letters home or sat silently--waiting. But, "there was an undercurrent of gayety, a little nervous, it is true," wrote historians, "but nevertheless bubbling over with rather grisly pictures of what would be done to the Germans." The slow pace of the passing hours felt like they had "leaden feet" as the huddled men contemplated the coming danger discussing tactics against machine guns, the effects of mustard gas of which lecturers had spread a healthy fear, before talk drifted to remembering home.

At 5:30 a.m. came the terse order to fix bayonets! At the ready, the soldiers grew quiet in tense apprehension. At five minutes to six o'clock the cry went out to attack! --hundreds of men pouring "over the top" sinking into the heavy fog which blanketed the area. Advance squads worked to clear broken barbed wire which luckily had been largely flattened by the heavy American guns. As men raced across no man's land, no shots eerily came from the enemy. The 362nd moved in slow advance into the wooded Bois

Chehemin until surprised by the lethal interruption of machine gun fire. Their determination, however, gained the division eight kilometers that day, gaining the village of Epinonville only to abruptly lose it to a German counter attack.

The Argonne Forest consisted of ravines, hillocks and small winding streams among thick underbrush. The small Aire River ran along the base of a slope to the west, its high ground giving cover to German artillery which had complete observation of the area. Dozens of German heavy guns to the east on similar rises and ridges nearing the larger Meuse River, rained down hellfire on the Americans.

Over the next two days the 362nd again advanced under a barrage of German artillery until reaching the reverse slope of a ridge, two and a half kilometers from the village of Gesnes just south of Romagne. Their goal on September 29th was that village and penetrating the Kriemhilde Stellung defensive line just beyond. It was the final and most formidable of the German defensive trenchworks.

The attack began at 7 a.m. with troops first merging through the wooded Bois de Cierges. They would then have to push through a shallow, open valley exposing themselves to the withering fire of the ridge tops on both sides. They were soon pinned down by machine guns and light artillery, bringing the attack to a halt. Regimental commander Colonel John Henry Parker, fearing annihilation, felt it was impossible to advance further. His superior disagreed. Parker received orders that Gesnes must be taken at all costs—an order Parker believed was suicidal. Following another American artillery bombardment, the 362nd rushed forward in waves starting at 3:40 in the afternoon and were hit with a "frightful volume" of enemy shot and shell at near point-blank range. The charge took two hours with the regiment being the first to break through the formidable Kriemhilde Stellung even as they left hundreds of their dead and wounded strewn across the bloody meadow they had just traversed. One of those bodies was that of a cannery worker from East Stanwood, Andrew Wik, who was reported cut down by machine gun fire.

Regimental historians recorded that only a handful of men escaped the slaughter. They had achieved their objectives, but at a terrible cost. Then came the unimaginable—a new order to abandon all that had been gained and retreat back to their original line of that morning. In their sacrifice they had outpaced the regiments on their flanks and were now in an indefensible position forcing their withdrawal. Regimental historians recalled the shock of those who were left:

"No one can describe the feelings of the men when they received the order and realized what it meant: that the ground which they had taken at such terrible cost was to be given up and that the blood of their comrades had been shed in vain. Each man felt bitterly that he had participated in a veritable 'charge of the Light Brigade'—heroic, perhaps, but futile…"

"As the men fell back in the darkness, they gathered up most of the wounded. The night was black and cold and rain that was half sleet fell. … The moaning cries of the wounded seemed to come from everywhere out of the darkness. Here and there a man was found wandering about among the dead and wounded like a lost child; rendered so by the terrible shock and horror of the carnage. How unutterable sad and heartrending, how awful is the aftermath of a war-swept battle field—and especially if fought in vain!"                    (The 362nd Infantry Association)

## Two Boys from Norman:

### Bert Stevens and Andy Engebretsen

Two of the soldiers who went "over the top" on the first day of the offensive were from the small farm community of Norman in Snohomish County—24-year-old Bert Stevens and 25-year-old Andy Engebretsen. Entrance into the war found both young men steadily working their father's respective farms.

Dark-haired Andy Engebretsen was born in Norman in 1893 but the stout, blue-eyed Bert Stevens arrived in the area in 1909 when 15, moving with his parents from his birthplace in Iowa. Both were on the Honor Roll of the newly expanded Norman school, where their Superintendent was one of the county's leading suffragettes, Mrs. Rainie A. Small. Bert also was on the Everett High School Honor Roll and attended Washington State College for a year. The boys together fulfilled the draft quota for the Norman community in 1917. Local residents used personal cars to transport the boys to Fort Lewis where they reported together on September 19, 1917. Four months later Andy married Madga Sofia Borgen. At Fort Lewis both were assigned to Company D, 361st Infantry, 91st Division. Andy was soon promoted to corporal. After arriving in France, Bert was transferred to Headquarters Company.

BERT STEVENS

The 361st Regiment was on the left flank of Wik's 362nd and both Stevens and Engebretsen were in the initial wave of the Meuse Argonne offensive on September 26th. As they moved across the desolate, killing fields during the first day's fighting, the 361st faced German artillery and machine-gun fire--light at first but that would change. American artillery threw mustard gas and phospene at the German lines and led American troops with a smoke barrage, despite the already heavy fog, leading to some initial confusion among the green troops. Still they persisted, moving German lines back eleven kilometers. Several members of the 361st, many from Company D, were awarded the Distinguished Service Cross for bravery that day including three from Headquarters Company for actions to repair telephone lines

CORPORAL ANDY ENGEBRETSEN

and secure communication with regimental command, all while exposed to enemy fire.

The confusion and carnage of the battlefield often makes it difficult to determine the exact time of a soldier's death. Records suggest that Stevens died on the first day while Engebretsen was killed later on October 11th just before the regiment was relieved that evening. A newspaper account printed in April 1919 offers rare insight into the reported fate of Stevens.

Stevens served under Headquarters Company Sergeant John Roman who was described as one of the best in the 91st Division. Officially, Stevens died on September 26th. Lieutenant Colin V. Dyment, however, recalled an episode that suggests Stevens died on September 27th. The account later ran in the *Oregonian* and the *Seattle Daily Times* on April 21, 1919. On September 26 the 91st Division breached a German defensive works called the Völker Stellung which brought them to the outskirts of the village of Epinonville which lay on the slope of a canyon. An orchard on its edge was brimming with German machine gun nests and snipers. Sgt. Roman identified one of those nests which had just wounded an American soldier. Taking the lead Roman surveilled the situation, then organized the crew of a one pounder gun. This was a small maneuverable French weapon which fired 37 mm shells and generally had a crew of two. It was primarily used against enemy machine guns. Bert Stevens was in the seven-man group that assisted Roman. He and Private Gustaf Peterson were ammunition carriers for the gunners.

While positioning the one-pounder, Roman was wounded in the back and Private Gerald Davison took over the squad. They maneuvered behind a hedge near the top of the canyon slope, but after two enemy shells hit close by the men scrambled to find cover in nearby foxholes. While moving the weapon a third shell landed among the crew. Several men were wounded but Dyment recorded that Peterson and Stevens "were blown to instant death."

The battle's first phase ended on October 1st. The depleted 361st marched eight miles through a driving rain and secured defensive positions, while under enemy fire, near Gesnes, France. On October 4th, American and French troops again attacked all along the front in another massive assault. The 361st Infantry and Andy Engebretsen led the way against fortified German hilltop 269. Although "severely handled" the objective was gained. Andy Engebretsen survived for another week before losing his life on October 11th. The AEF lost just over 26,000 men in the Meuse-Argonne Offensive. Six days of fighting had cost the 361st 1,257 casualties. Nearly 300 of those were killed in action, including two farm boys from Norman.

# Stanwood's Symbolic Soldier
## Frank Hancock

Many people have seen his slightly faded name printed on the cream stucco south wall of American Legion Post 92 in Stanwood. The Post is named in his honor—Frank H. Hancock. The young man from East Stanwood was only 26 when he died in the Battle of Meuse-Argonne; one of over 116,000 Americans who gave their lives in World War I. The local farm boy was drafted in June 1918. He would die three months later.

Francis H. "Frank" Hancock was born in East Stanwood on February 25, 1892. He was named after his grandfather, one of the early pioneers to the Stillaguamish River valley. Francis senior left Missouri for the Northwest in May 1862 to escape the growing

*Frank Hancock*
*Photo courtesy of Stanwood Area Historical*
*Society 87.10.56.09*

conflict of the American Civil War. A few years later, he moved his family from Whidbey Island to farmland just north of Centerville (today's Stanwood). Young Frank Hancock worked the land with his father John Trader Hancock until called for service.

Frank trained as a rifleman at Fort Lewis, Washington and Camp Kearney, California before he shipped out as a member of Company C of the 158th Infantry arriving in France in August 1918. The 158th did not see action; its men used as replacements for other units. Frank was quickly transferred to Company K of the 110th Infantry, part of the famed 28th "Keystone" Division, comprised mainly of troops from Pennsylvania. The 110th had taken heavy losses in July and August supporting French troops in the Second Battle of the Marne. It would again be a lead unit in the largest assault of the war, the Meuse-Argonne offensive, which began September 26, 1918, where the regiment sustained losses of over 50%. Their advance in the Aire River valley would capture several key cities on the 26th including the town of Varennes, a victory which collapsed the German lines.

Paul Ross, said to be Hancock's "comrade," wrote his parents that Frank died that opening day of the attack. Ross said he later found Frank's body "rifle gripped in his hands," having fallen facing the enemy with the expression of anger still frozen on his face even in death. He died instantly, Ross assured his parents, as evidenced by the bullet hole which entered his side. This story, however, conflicts with the official explanation given by his regiment, based on the eyewitness account of a soldier who was with him.

In that 1920 regimental history, Corporal Edward Smith, also of Company K, reported that "I was wounded by the same shell that killed Frank Hancock. We were members of a mopping-up party and were walking along the road in the town of Varennes. The shell hit about ten feet from us and for a few minutes I was stunned and when I recovered I found Hancock dead. He was hit in the head by a piece of shrapnel and must have died instantly." Smith stated that Frank died two days after Ross' account, on September 28th. Frank Hancock was later returned to the U.S. and buried in Arlington National Cemetery where his headstone bears the date of September 26, 1918 as the day of his death. However, it also had his home as the District of Columbia instead of the state of Washington; a mistake corrected in 2021. Clearly in the chaos of war and its aftermath, mistakes are made.

Frank's mother Annie Astel Hancock was one of the first Gold Star mothers to accept the government's offer to pay their travel expenses to Europe in order to visit the grave of lost loved ones. She was also a president of the American Legion Auxiliary in Stanwood.

The reason for the disparity in the stories of how Frank Hancock died is unknown but it is not beyond the realm of possibility that facts were altered in the letter to Frank's parents to perhaps soften the blow of his death. There is no listing of a Paul Ross belonging to the 110th regiment but other units were in the area during that part of the offensive. Regardless of how he died, Frank Hancock's death in the service of his country places him on a hallowed list of those we remember on the Community Veterans' Memorial at the Stanwood Area Historical Society. The name of this symbolic soldier of Stanwood is now etched in marble.

# Willing to Pay the Price

## Joseph Bruseth

As bloody as the first phase of the Meuse-Argonne battle was, many believe that the next stage was harder. That began October 4th and would last ten days. Joseph Bruseth was also a member of the 28th Division in a sister regiment to Frank Hancock, Company L, 111th Infantry. The division had extended the American lines to Apremont by September 28th but another fortification was an imposing threat—Le Chêne Tondu. Historian Thomas Fleming described it as "a saw-toothed height jutting out of the Argonne Forest into the valley like the prow of a ship." The ridge dominated open ground to the east in the Aire River valley. The rocky outcropping along the west plateau of the Argonne Forest would not be taken until October 7th.

Eliza Bruseth could not accept the news that came to her door by telegram in December 1918: "…it is officially reported that private Joseph E. Bruseth died October fourth from wounds received in action." Joseph was "instantly killed on hill Chêne Tondu … by machine gun fire." Eliza and Joseph were born just two years apart in a family of eleven children. She began a campaign to uncover more details about her brother's death.

Joseph's parents were John and Ellen Bruseth. John, a Norwegian immigrant, farmed in both Skagit and Snohomish Counties from the 1880s to the early 1900s. Joseph was born in Silvana in May 1890. Joseph's Uncle Nels was a well-known forest ranger and amateur anthropologist, geologist and musician in Snohomish County. Personal difficulties may have intruded on the family, separating them by 1910. John and son Gundar remained in Skagit while wife Ellen ran a farm in Elk City, Oregon with Joseph and her other children. However, by the time Joseph registered for the draft in June 1917, he is working for the Florence Logging Company in Snohomish County.

*Joseph Bruseth*
*Photo courtesy of the family*

Eliza's persistence in seeking answers about her brother's death resulted in a letter dated just over a year after Joseph was killed. It was from Company L's First Sergeant Albert P. Schad. Schad recalled Joseph and the private's unsuccessful attempt to transfer to Shad's platoon and his part in the regiment's efforts to secure the fortified ridge of Le Chêne Tondu from the Germans. Americans achieved the crest of the ridge on October 1st and 2nd only to be halted in front of German trench lines on the hill's rear slope. Their assaults "came to grief every time," wrote historian Edward Lengel, forcing their withdrawal. The mounting casualties greatly undermined American morale with one doughboy writing that "above our heads bullets snap and crackle. We talk of home and God. The terror of the night is upon us. We realize the hopelessness and uselessness of it."

Schad told Eliza that with another assault planned for October 4th, Joseph, as part of a four-man patrol from First Platoon, was sent into no man's land to reconnoiter enemy strength. Five minutes after they began, two of the soldiers returned with their wounded corporal who, before he died, informed Schad that Private Bruseth's dead body remained on the open field. Schad would find Bruseth the next day, after leading a successful assault on the ridge for which Schad was awarded the Distinguished Service Cross. Upon examination, he confirmed that Bruseth had been shot through the left lung—one exploding bullet sending shrapnel though his body. We don't know if the information eased Eliza Bruseth's mind, as Sergeant Schad had hoped. The family later had Joseph's body returned to the U.S. and buried in Arlington National Cemetery.

General John Pershing's tactics of open warfare have found criticism in later years and he was accused of underestimating the lethality of the improved machine guns of World War I. At war's end, on armistice night, Pershing stood to offer a toast: "To the men," he said. "They were willing to pay the price."

# Argonne Hellscape

## Alfred Kristoferson

Alfred Kristoferson, Jr. was born a year after his father and namesake moved to Washington state. He was destined to be heir apparent to his father's very thriving dairy business in Seattle which Alfred Kristoferson, Sr. began in 1896 after immigrating to America in 1881 from Sweden. His oldest son was one of four children in the household of Alfred Sr., and his wife Marcia Alberta Clarke Kristoferson.

Alfred Kristoferson Sr.'s untimely death in March 1914 at the age of 55, thrust his 23-year-old son into a management position earlier than anticipated. His father had found opportunities in America, beginning with farm work in Momence, Illinois just southeast of Joliet. He later opened a mercantile and married Momence resident Marcia Alberta Clarke September 9, 1886. The family moved to Seattle, Washington in 1890, but relocated to Mount Vernon in Skagit County the next year. In 1893 the family bought some land around Florence, Washington five miles south of Stanwood and began farming. Five years later, however, the family was back in Seattle where Alfred purchased ten acres on Mercer Island. A year later the Kristoferson Dairy began.

Deliveries were often made by boat around Lake Washington. The business continued to grow with operations expanding into Seattle proper. In 1912 Alfred Sr. purchased several acres of land on Camano Island for a dairy farm, intending to give that to his second son August. His oldest son's death would change those expectations although Alfred Sr. did not live to hear of his namesake's death, dying in 1914 in Hollywood, California where he spent several months trying to improve his failing health. He was surrounded by his family at the end including Alfred Jr. He had been involved in a serious auto accident two years earlier which may have contributed to his decline. At the time of his death, Kristoferson Dairy had grown to become the largest company of its kind in Seattle.

Alfred Jr. had interests outside of the dairy business that included military service, holding the rank of private in Company L of the Washington National Guard. In June 1916 the Washington state troops and other state units were mustered into service for three months to patrol along the Mexican border from their main camp at Calexico, California. Their three-month job of border protection came at the time that General John Pershing was attempting unsuccessfully to capture Mexican revolutionary Pancho Villa. Alfred Jr. would have a short respite but again

*Alfred Kristoferson Sr.*
*Photo courtesy of the family*

found himself called to action at the onset of America's involvement in the world war. His 2nd Infantry Regiment of the guard was federalized on March 25, 1917 and became part of the 161st Infantry Regiment, 41st Division, twelve days before the United States declared war on Germany.

The Division, and now Corporal Kristoferson, shipped to France with the regiment's first units in December of 1917; its last units going in early February 1918. The 41st did not see the front lines after arrival in France, instead becoming the 1st Depot Division; its men often used as replacements for other units. Corporal Kristoferson reportedly was more involved in construction details and training exercises which did not set well with him. His request for a transfer to a combat unit was finally granted and he became part of the 32nd Infantry (Red Arrow) Division comprised of Midwestern men mainly from the Michigan and Wisconsin National Guards. The transfer also resulted in another stripe. Sergeant Kristoferson was in Company A of the Division's 126th Infantry Regiment.

The 32nd and 91st Divisions would be in the forefront of the Meuse-Argonne offensive. Both divisions were deployed in V

Corps. Andrew Wik died during the first stage of attacks on September 29th; Alfred Kristoferson followed him during the offensive's second phase on October 15th. The 32nd division had been in reserve during this phase of the battle and would relieve the battered units of Wik's 91st Division.

The goal of both divisions was to break through the fortified line of German defenses termed collectively the Hindenberg Line—their specific target point called the Kriemhilde Stellung, the strongest position of the German line. Lt. General Robert Bullard, commander of the 2nd U.S. Army, described it this way: "The way out is forward, through the Kriemhilde Stellung, eastern section of the Hindenburg Line. ... Not a line, a net, four kilometers deep. Wire, interlaced, knee-high, in grass. Wire, tangled devilishly in forests. ... Pill boxes, in succession, one covering another. No 'fox hole' cover for gunners here, but concrete, masonry. Bits of trenches. More wire. A few light guns. ... Defense in depth. Eventually, the main trenches. Many of them, in baffling irregularity, so that the attacker cannot know when he has mopped up. ... Farther back, again defense in depth, a wide band of artillery implacements." It would prove a killing field for a lot of American troops.

Those same fields would have to be crossed again by the 32nd Division which relieved Wik's shattered 91st Division on October 3rd. The following morning began the second phase of the Meuse-Argonne offensive. Ironically, the 32nd, as with Kistoferson's original division, was initially assigned as a supply and training division but its status was changed to a combat role April 10, 1918 and it moved to the Alsace region where they were assigned to the French 40th Corps. We don't know the exact date of Alfred's transfer but certainly after this point. At Alsace in late May the division suffered its first casualties in sporadic firefights. On July 19th they were pulled and moved to Château Thierry and endured their first true baptism of fire on July 30, 1918. They would suffer over 4,000 casualties before being relieved on August 7th.

They became part of the Oise-Aisne offensive on September 5th although now again under American leadership. Another 2,000 men fell during this time. The 32nd Division received the highest order of the Croix de Guerre; the only National Guard unit so honored during WWI. General Charles Mangin said that the Division, "proved its superiority in a fierce hand-to-hand struggle where the 125th and 126th Regiments emerged victoriously despite counter-attacks by the enemy."

The Division received a well-deserved rest on September 10, 1918 not realizing that another severe test of their stamina and courage would follow very soon. Nearly 5,000 green replacements joined the ranks of the 32nd to replenish the Division's companies. In late September the Division was re-deployed to the lines behind V Corps as a reserve unit during the opening of the Meuse-Argonne offensive. On the day Andrew Wik died, September 29, Sergeant Kristoferson and the 126th Regiment along with the rest of the 32nd Division, pulled on 78-pound packs and marched eleven miles on a chilly, dark and rainy night across a shell pocked landscape, littered with shards of wire and broken trees, the signs of earlier battles. Over the next early, cold days of October they occupied the trenches which had held the divisions of V Corps and prepared for their own assault on the fortified Kriemhilde Stellung.

Their hell in the Argonne began October 4, 1918. This time the goal was the village of Romagne five miles north of the village of Gesnes, bracketed by the forested and fortified hills of the Côte Dame Marie and the Côte de Châtillon—the former being the key to breaking through the Kriemhilde Stellung. Fighting over the next few days would see Gesnes taken and relinquished as pockets within the Argonne were cleaned, with great efforts by the Americans, of garrisoned and well-hidden German machine gun nests. By October 8th the Americans had pushed their line roughly two miles north of Gesnes. Following another intense barrage by American artillery, the division pushed to the outskirts of Romagne on October 8th engaging in fierce hand-to-hand combat in the deep, fortified and irregular trench system. After one day of reprieve away from the front on October 12th the regiment was ordered back just in time for another major push to penetrate the German Kriemhilde Stellung.

The dawn attack on October 14 again started with American heavy guns blanketing the German lines with shell. Troops then charged through the tangle of wreckage with the 126th having the best luck in its rush forward. With other regiments bloodied by the withering machine gun fire from the concrete and masonry protections on the heights of Côte Dame Marie, the 126th was able to push past the deadly hill and penetrate the seemingly intractable Stellung. This allowed them to flank the Germans on the merciless incline, forcing their surrender. With this costly success, another assault was planned for 7 a.m. the next morning. Fighting that day consolidated the American line and completed the breaching of the Kriemhilde Stellung. Records say that October 15th was the last day of Sgt. Alfred Kristoferson's life. He is not mentioned in the regimental history but records indicate that only one man from Company A, was killed that day. His death date may also be speculative since he was initially listed as missing in action suggesting that his body was not recovered until later on the desolate battlefield. By early January 1919 the *Seattle Star* reported the official announcement of his death.

Both Andrew Wik and Alfred Kristoferson are buried a few rows apart in the Meuse Argonne American Cemetery in Romagne, France. We know of no sentiments or elegies for Private Wik. However, Alfred's uncle by marriage, Kiyoshi Kawakami dedicated his book *Japan and World Peace* to Alfred when published in 1919. Kawakami, a respected Japanese journalist and writer of the time, was known as K. K. Kawakami after taking the middle name of Karl after Karl Marx and, although based primarily in San Francisco, was a co-organizer of the Socialist Party of Seattle.

Marcia Kristoferson accepted the government's offer to widows for gold star mothers to visit their loved one's gravesite in France, sailing on the S.S. *Republic* in the summer of 1931. A 1928 newspaper article said of the program: "This is a trip that many of the bereaved have longed for and would have taken but for the hard obstacle of cost. The expense which the government is to bear will be small compared with the satisfaction of sentiment afforded by these visits."

# An Immigrant for America

## Jacob Teiseth

Jacob Teiseth's physical appearance may not have been what many would imagine for a war hero. His World War I registration papers describe him as "stout" at 140 pounds and only five foot three inches tall. His time in America was also short before being called to defend his adopted country. Emigrating in June 1913 aboard the ship *Kristianiafjord*, he was still a citizen of Norway when he entered the U.S. Army in 1917, listing his new hometown as Stanwood. He was assigned to the 6th Engineer Regiment of the 3rd Division which were some of the first troops to see battle in Europe. Jacob was one of the regiment's field medics.

Jacob Bernhard Teiseth was born in Oxendalin, Norway on May 13, 1894. He followed his older sister Gudrun to America. Like so many Norwegians before him, Teiseth found employment in the saw mills—in his case the shingle mill of John Andall near South Bend, Washington on the coast. He wrote that he was the sole support for his widowed mother back in Norway.

Gudrun had emigrated in 1910 and by 1917 was working as a domestic servant in Seattle. That same year she married widower Lars Husby, whose Norwegian hometown was the same as the Teiseths. Starting as a farm laborer, Lars had acquired his own farm along the Stillaguamish by 1917. Jacob signed as one of the witnesses at their wedding, just two months before leaving for France.

*Jacob Teiseth*

Three days prior to Kristoferson's 32nd Division relieving the 91st, the 3rd Division and Teiseth's engineers relieved V Corps' 79th Division just to the east of where the 32nd had settled in. Reports indicate that the 6th Engineers got little rest between April 1918 and the Armistice the following November. Unfortunately, Jacob would not live to see that day. During the offensive's second phase, the 6th Engineers were assigned to assist in capturing a triangular forest known as Claire Chênes Woods which lay just north of Brieulles along the Meuse River. The front line and two hills on the wood's flanks bristled with upwards of 200 machine-gun emplacements making any action deadly work for the Americans.

An attack early on October 20, 1918, however, proved successful in driving the Germans back and gaining the hills and the forest, but at a high cost against fierce enemy fire. Captain Charles Harris was shot through the lungs as he and 12 men turned captured machine guns on the enemy. A German counterattack overwhelmed the now diminished unit which lacked any support from other units supposed to protect its flanks, and they were forced to relinquish their hard-earned ground. Rallied again, the regiment re-took the Claire Chênes and would hold it for five days until relieved. The regiment lost 27 men that day, with another 114 wounded.

Several men were awarded the Army's Distinguished Service Cross for gallantry during the battle including Captain Harris and medic Jacob Teiseth. The citation tells that Private Teiseth worked throughout the attack under constant machine gun fire removing wounded men and supervising their evacuation from the battlefield before being cut down himself. The *Stanwood Tidings* newspaper reported that Teiseth's bravery was brought to the attention of General John Pershing who awarded the young Norwegian his medal posthumously upon the recommendation of Teiseth's commanding officers. He had brought Stanwood into "favorable limelight" through his heroic work, wrote the *Tidings*. The young emigrant was back in Europe again, this time in the vast Meuse-Argonne cemetery, but, added the paper, his "deeds will live forever."

# Laid in a Soldier's grave.

## Albert Buli and Ray Bunton

The voyage on the S. S. *America* took nine days. As Margaret Buli and her fellow Gold Star mothers and widows were ferried to the pier at Cherbourg, France, they were greeted by excited smiles, waving flags, and colorfully festooned boats in the harbor. Waiting onshore was a welcoming committee of enthusiastic local officials and French women who had also lost loved ones in the Great War. It was the spring of 1930 and Buli and her fellow travelers were an early contingent who had accepted the American government's offer to visit the graves of their lost sons and husbands. Margaret had traveled from Pierce County, Washington. Nearly 7,000 Gold Star Mothers and widows would make the pilgrimages over the next three years. Their average age was 65 with the oldest traveler 91. Margaret was 67.

The mothers, in particular, were described as "soberly dressed, white-haired women." They were well cared for. The federal government appropriated over 5-million dollars for the pilgrimages. All reasonable expenses were paid by the government which also provided an Army officer escort and a nurse. Shuttled to Paris by special boat train, they would be taken to see the gravesites at the specific cemeteries holding their family members. Their biggest fear was that the sites of internment would not be properly maintained or simply ignored. Their fears were soon alleviated.

For Margaret Buli the vast Meuse-Argonne American cemetery was her desired destination. She wished to offer a final goodbye to her son Albert Noren Buli who rested there. The young 24-year-old Marine private died during the last great Allied offensive of the war—his death coming just five days before the armistice which ended the bloodshed.

*Gold Star Mothers on the S.S. America*

Private Ray Bunton of Warm Beach died the same day as Albert Buli but under much different circumstances. Some people seem to just fade from the pages of history or suffer the fate of not having their lives recorded. For Ray Bunton, there are few footprints. Bunton's family migrated from Carter County, Tennessee where Ray was born June 18, 1897 at Midvale, Idaho

*The hospital at Fort Worden*

where his father James R. Bunton farmed. The red-headed, blue-eyed Ray registered for the draft while working in Almira, Washington as a railroad worker on August 24, 1918. That may be why by 1930 the family had moved again to Washington state and settled in the community of Warm Beach tucked alongside Port Susan.

After being drafted, Ray was sent to Fort Worden near Port Townsend for training. Fort Worden, part of the so-called "triangle of fire" along with Forts Casey and Flagler, was built to protect the Puget Sound. It was first garrisoned in 1902 but greatly expanded after the U.S. entrance into the war, adding six new buildings. It was not long, however, before Ray was hospitalized in the camp hospital where he died on November 6, 1918. We don't know the cause of death but the fort was dealing with multiple cases of influenza at that time. Eight had died with 200 ill by late October and one newspaper described the hospital as "crowded" with patients by late November. It is likely that Ray was a victim of the silent enemy in 1918.

His body was taken back to Midvale, Idaho for burial and a large gravestone was erected over his remains. It lists the private's service number and the designation of "40th Co C ACPS." Below it reads "Gone but not forgotten." A sad irony on his life since his last name is misspelled and the gravestone carries an inaccurate birthdate for the young soldier. Lastly, his epitaph is a poem chiseled on the bottom of the stone:

"He left his home in perfect health

He looked so young and brave

We little thought how soon he'd be

Laid in a Soldier's grave."

Albert Buli also died on November 6, 1918. At the time of Albert's enlistment the family was living in the Cedarhome section of East Stanwood, Washington. He enlisted early in the war, picking the iconic American holiday of July 4, 1917 to join the Marine Corps. The family patriarch Ole and his sons, including Albert, ran a dairy operation. Originally from Iowa, where Albert was born in July 1894, the Buli family had not lived long in Stanwood, moving there sometime after 1910. Albert was one of nine children in the Buli household. His Norwegian heritage was evident in his light blue eyes and sand colored hair. Albert's brother Thomas Buli was in the same company of the 361st as Andrew Engebretsen. He was wounded in the left hand by machine gun fire on September 28th in the same action that took the lives of Bert Stevens and Andrew Wik. Thomas survived the war.

ALBERT BULI

Albert trained on Mare Island in California and was assigned to the 5th Marine Regiment of the 2nd Marine Division. He was part of the final battle of the final phase of the Meuse-Argonne offensive which pushed the Germans to surrender in World War I. It was the fourth major offensive for his division. The greatest challenge of this last push of the war fell on the American III Corps and Marines within it. The Corps was positioned on the far right flank of the American line, the French XVII Corps just to the east of them with the Meuse River between.

The so-called Red Devil Dogs won praise for a tenacious assault beginning on November 1. They pushed the Germans back nine kilometers on the first day and destroyed the last stronghold on the fortified Hindenburg Line—the Freya Stellung. The unrelenting progress of III Corps succeeded on the evening of November 4-5 of driving the Germans north back over the river. Albert Buli is said to have died during this last portion of the offensive—five days before the end of the war.

Margaret Buli left France aboard the S. S. *America* on June 19, 1930. The culture of France had seemed strange to many of the American mothers, used to a simpler and often more austere life. It's not known if she learned anything about the specific circumstances of her son's death in that foreign land. She surely felt that her son was a long way from the lush, quiet fields and forests of Washington. However, the manicured, green lawns and peaceful beauty of the cemeteries must have demonstrated the reverence and esteem that the people of France held for Albert Buli and all those who lay with him so far from home.

# As If God Was Calling Them

## Emma Thorsen

*Nurse Emma Thorsen*

By mid-April 1918 there was a heightened level of concern among medical officers at Camp Dodge, Iowa. Soldiers at this central training facility were dying at an ever-alarming rate and they weren't sure why. Fifteen soldiers had died in less than a week of pneumonia prompting an investigation. The fatality rate was 30 percent for those soldiers affected. Although reports existed of influenza throughout the Midwest in December of 1917, staff at Camp Dodge thought the more likely source of their problem was the vigorous drill duty of the troops while being exposed to the thick dust clouds that swirled through the encampment. Camp Dodge was a central post for the training of army troops heading for the war in France—both African American and Anglo. Before the year was over the demand for nurses became desperate. Seven of them would die at Camp Dodge, including a 28-year-old Red Cross nurse from Silvana, Washington.

Born on August 30, 1890, Emma Josephine Thorsen grew up on a dairy farm. Her Norwegian father, Halvor, came to the United States in 1887 where he found work in the mines of Michigan and Colorado and finally the logging camps along the Stillaguamish River in Washington where he may have developed his support of the Socialist Party cause. After a short adventure in Alaska, he returned to the area and began farming. Emma's younger brother Carl remembered the raw landscape where he, Emma and their siblings grew up with few roads. School was a four mile walk along narrow trails through hollows and hills to nearby Bryant and later the Silvana schoolhouse.

Emma worked as a servant in Everett while studying to become a nurse at Providence Hospital, graduating in 1915 alongside fellow classmate Mayme Downs. We don't know why Emma decided to join the nursing corps but we do know that her assignment to Camp Dodge came at the height of the influenza epidemic—like a storm, one writer remembered. Nearly 14,000 personnel would be hospitalized for influenza at the Iowa camp which prepared men for combat overseas. According to science historian Carol Byerly the Iowa facility "had one of the worst records among Army camps" for infections and deaths.

Thorsen, Red Cross nurse Irene Robb and the other nurses were part of roughly an additional 400 nurses who were sent to Camp Dodge to help with the outbreak, working 12 to 14-hour days. Robb wrote of the difficult time. The nurses worked in hallways and wards crowded with the ill and dying without protective clothing, adequate procedures or any medication or vaccines. According to one group of researchers, "in spite of their technological poverty, nurses and physicians stuck to their posts in the face of the most lethal medical disaster in history."

Robb's letter home speaks to the exhaustion of tending for so many without sufficient help; some nurses responsible for 150 patients at a time. Robb said that when it was all over, she would simply collapse until she recovered from all the sadness she had witnessed. Her patients were "grand men" and so appreciative of all that was done for them, "which was really ...very little," she wrote. "It seems just as if God was walking up & down the ward just calling those he wants & they just all go," she wrote family members.

Pneumonia took the life of Private Wayne Loveless on January 22nd. In early February a black officer reportedly died of the lung ailment, followed a few days later on February 18th by two more recruits—one a Choctaw Indian, the other white. This was no isolated incident among army camps in the country. At Camp Cody in New Mexico ninety percent of deaths since November 1917 were attributed to pneumonia. That figure was under 50 percent for Camp Dodge. Some studies point to China as the source of the original outbreak and even Kansas is seen as a possible early center since 48 soldiers died at Camp Funston in that state in March 1918 at the same time pneumonia was taking lives at Camp Dodge. Ironically, an article in the February 18th *Evening times-Republican* of Marshall, Iowa lauded the sanitary conditions at the Iowa camp. "Cleanest camp in the country," a writer for *Collier's* magazine reported after visiting. It was in line with the camp's reputation. The Secretary of the State Board of Health declared the camp a wonderfully healthy place—joking that the only place where that was questionable was the camp rifle range.

By the latter days of April 1918 almost two thousand men at Camp Dodge were hospitalized with as many as 32 a week succumbing to the assault on their pulmonary system; the largest death rate at any cantonment in the country. Men believed to be unfit for the rigors of army life were being weeded out and discharged. Deaths from the disease reportedly dipped for a time in May by 10 percent with the army's surgeon general proclaiming that general health conditions in the camps were "very good." June reports declared that the problem was "rapidly decreasing" from 112 deaths in April to only 55 in May of 1918.

Pneumonia seemed to vanish from the headlines of local Iowa newspapers in July and August with the downturn in cases. Camp Dodge again focused on training the thousands which were being readied for the European war. To rally public sentiment and sell bonds, soldiers gathered and formed human statues of iconic American symbols in the summer of 1918; everything from a bust of President Wilson to the Liberty Bell and divisional patches. Specially built towers were built to aid in organizing and photographing this human spectacle. At Camp Dodge 18,000 troopers gathered on the drill ground on August 22, 1918 to create the "Goddess of Liberty" which stretched for a quarter of a mile—"a living representation of Liberty Enlightening the World," as described by the *Bottineau* [ND] *Courant*. Dressed in woolen uniforms, twelve men fainted in the 90 plus degree heat.

A wave of warm air persisted through the sultry beginning of September in Iowa. The first publicly identified case of influenza at Camp Dodge was said to be September 13, however a 2011 study of autopsy samples showed the first verified influenza fatality was a soldier at Camp Dodge on May 11, 1918. This faceless enemy, which had never really left, struck with a vengeance by September 29th. The next day a small paragraph in the *Evening times-Republican* announced "Camp Dodge Invaded" by influenza and a quarantine was expected. But the article only made it to page eight of the paper. On the same page, alone amid stories of livestock prices, an article warned that flu was expected to "grip half of Nation," but the story downplayed the risk. While as much as 60 percent could be infected, the vast number of people were expected to recover according to officials.

Over six thousand soldiers were ill in the first week of October overwhelming Camp Dodge's medical services. The *Des Moines Register's* Frank Santiago wrote that camp documents showed that "many soldiers who awoke healthy were sick by noon. They were dead before supper. The stunning speed of death left the camp reeling." Division surgeon Lt. Colonel E. W. Rich stopped issuing daily reports saying they were causing "unnecessary worry." Documents would later show that 100 men died the first week, 350 the following week and 450 by mid-October, surpassing what would end up being the official count.

Lt. Colonel Rich lifted the quarantine put in place at Camp Dodge by October 27th. The *Denison Review* reported a great relief in the camp as soldiers celebrated with "parades, band concerts [and] resumption of all kinds of amusements and the mingling together as in the past." The "terrible visitation" had reportedly left as fast as it had come. A month later it was reported that a little over 10,000 influenza cases resulted in almost 2,000 cases of pneumonia leading to 702 deaths. Historians, however, place the death toll at around 1,000. It was reported that the majority of the victims were black; three to one according to a study published in 1919 by the assistant camp epidemiologist.

No evidence could be found, however, that the deaths from pneumonia at Camp Dodge in the spring of 1918 were included in the final tally of influenza deaths. Of the seven nurses listed as dying during the epidemic at Camp Dodge, Everett resident Mayme Downs died October 24th followed a month later on November 24th by Emma J. Thorsen of Silvana township. A Red Cross history noted that "Red Cross nurses and nurses not enrolled, nurses available for service later on and nurses who would never be eligible for permanent enrollment, packed their kits, boarded the trains and proceeded like soldiers to the camps." Nurse Robb remembered that the soldiers died as if God was calling them. Nurses heard that call as well.

## "The Calling"

Do you sometimes wonder?
Why you do the job you do?
Did you choose your career?
Or did your job choose you?

Way back before you were born.
God knew there was a need.
So He picked your fertile heart.
And planted a caring seed.

Then He waited and He watched.
Knowing before too long.
The desire in you to help others.
Would continue to grow strong.

The caring heart He put in you.
As you put others first.
Leaving only one path to take.
In you there was a thirst.

Not seeking fame or fortune.
Born with a tender touch.
You are who you're meant to be.
That's why you care so much.

Because caring don't take practice.
It's not something you rehearse.
You answered a special calling.
When you became a nurse.

by Edwin C. Hofert

# World War II
## The War to Defeat Fascism

| | |
|---|---|
| Arne O. Aalbu | Gordon L. Lord |
| Robert G. Baker | Harold McCann |
| Roger Barney | Vivien D. Mickel |
| Robert S. Bransmo | Ernest C. Moser |
| Leonard M. Broin | Robert K. Nelson |
| Donald H. Garrison | Edward H. Pearson |
| Robert F. Harrison | Floyd D. Perin, Jr. |
| Daniel S. Hess | Peter T. Rekdal |
| Richard N. Hiday | Andrew W. Riker |
| Charles H. Isham | Dorman N. Riker |
| Orville P. Knutson | Wesley D. Sigerstad |
| Gerhard A. Lane | Alvin G. Vaara |
| Donald C. Leach | |

Those Americans who bore the burden of World War II have been dubbed the "greatest generation." Attributes assigned to this group of Americans, born between 1901 and 1924, are qualities such as integrity, personal responsibility, work ethic, frugality and sacrifice. The attack on Pearl Harbor galvanized this generation toward a common goal—defeating fascism. The residents of Stanwood, East Stanwood and Camano Island responded unwaveringly to the call of their country. Bond drives, food sharing and victory garden programs were eagerly organized.

Very quickly meetings were held to define and implement measures for defense especially for a town along the western coast of the United States. Local headlines declared: "BE PREPARED. We are in the Danger Zone." There was no shortage of volunteers. Air raid observation posts were designated and manned around the clock. Particularly in those first anxious days of the conflict, eyes scanned the heavens for strange aircraft or even balloons which might carry dangerous gases or threats of igniting the forests of the Northwest.

And Stanwood got busy. In December 1942 the Edlund Shipyard was established along the Stillaguamish River where once was housed the Stanwood Lumber Company and the Camano Blue Oyster Company. The first massive barge, delivered under a government contract, was ceremoniously plunged into the waters on July 22, 1943 as bands played and jubilant crowds cheered and waved flags. Morale was high as the newspapers proclaimed that "Shipbuilding will Boom Twin Cities." The ocean-going barges were loaded with supplies and equipment and towed to facilities along the west coast and Alaska. Shortly before the first launch, Edlund sold the operation to Horace Kelsey who renamed the business Stanwood Shipyards. The seventh and last barge was launched in August 1944. With the end to government contracts, Kelsey announced the cessation of shipyard operations.

The *Twin City News* often printed the words of the boys on the front lines, some in great detail, giving depth to the difficulties they were living through or tragically, what they were dying for. The men and women of the greater Stanwood and Camano area served in the far corners of the globe such as Dr. Harold Greer of the Army Medical Corps who got a surprise while relaxing at a base movie night. There on the screen the words flashed "Strange As It May Seem,"--a nod to the shortest steam railroad in the world—Stanwood's H & H. Images showed the Dinky engine chugging along through East Stanwood's main street with Mrs. Hall and son Jesse in the cab.

Lillian Barnes worked in the Edlund yards in 1943. Her husband John was a newspaper correspondent covering the war in Burma along the Chinese border. After going through his first air raid where the Japanese bombed their airstrip he wrote: "They came right in over our heads and unloaded about 150 yards from where we were standing but no one got scratched. Some fun!" Later he got the chance to catch a ride on a bomber and view the war from a few thousand feet. "'A little left, boy, now a little right,'" instructed the navigator, "'whoa, steady, steady now.' Then in a minute, 'Bombs Away' and see little messengers of destruction stream down...you follow them all the way down and then whoom! There is a big puff of smoke and flame and some little quiet spot has just departed for parts unknown with a few hundred pounds of TNT shovin' it along."

First Lieutenant Clarence Gansberg was a B-29 pilot stationed on Saipan in December of 1944. Resting after chow he heard the shooting from some enemy fighters strafing his airfield. Grabbing his helmet, he ran for the shelter and when almost there "looked up and saw one [fighter] coming real low across our area with his guns emitting vivid flashes of flame. I could hear the lead bees all around me so thought a good place to be right then was on the ground. Naturally, I hit it pretty fast! He went over and luckily missed me. What a sigh of relief I let out! I could just feel the bullet going into my back, but, thank God, it never came." Gansberg was awarded the Distinguished Flying Cross with clusters for his service.

Sailor Irving Utgard was aboard the ship *Ozark* at the end of the war in the Pacific. They left Guam on August 13, 1945 and "saw the B-29's pass overhead on their last hostile run to Japan and soon after received word that Hirohito had accepted unconditional surrender." He wrote his parents from Tokyo Bay. As they anchored they "saw several Jap submarines, each flying the black flag of surrender from its conning tower....It has been an honor to participate in this final amphibious landing and more so to be able to help evacuate our recently freed prisoners of war. May this begin a lasting peace!" Sadly, that was not to be.

# A Seeker of Wings
## Alvin Vaara

Alvin Gordon Vaara hoped to lift his young life to the skies as a pilot for the U.S. Army Air Corps. He enlisted on March 25, 1941, nine months before the Japanese attack on Pearl Harbor. The 23-year-old was with the first cadets to be sent to the Waco Army Airfield in Texas as part of the 54th Basic Flying Training camp in the spring of 1942. This was a change from the hard laboring life of his parents who had married in Seattle in 1911 before settling at Milltown, just north of East Stanwood. Alvin listed his hometown, however, as East Stanwood. His father Hans Gustav-- "Gus"--had been a teamster, farmer and shingle mill worker before becoming a fisherman in 1940.

In Waco, the cadets were the center of attention especially for the area's young women. Margaret Allison told a Waco reporter that "you could spot an aviation cadet a mile off. Every girl wanted to date them." The Cadet Club was a popular hangout. "There was no feeling like it in Waco," Allison remembered: "A room of cadets and their dates, enjoying the music of the day and a break from danger and thoughts thereof. Walking in that door, you were on the arm of a very brave young man who was risking his life everyday."

Dangers can be all around, however, even for lighthearted, youthful celebrants on a Saturday night. Such was the fate of Alvin Vaara in the dark hours just after midnight on January 24, 1943. Vaara and a new Pfc. recruit, John Richard Grzybowski, shared a taxi with a young married woman, 28-year old Lauretta Johnson. The cab was struck by a freight train as it negotiated railroad tracks at a grade crossing. All three passengers and the driver were killed instantly. Alvin now lies with his family in the Scandinavian Cemetery at Milltown. The wings Alvin sought would have to be earned elsewhere.

# A Soldier's Death

## Robert Baker

The announcement in a Seattle newspaper of the 31-year-old soldier's death was a sterile listing of basic facts: "Freeland, Island County. Staff Sergeant Robert G. Baker. Army. Europe. dead." A German sniper took his life on March 16, 1945 as troops of George Patton's Third Army penetrated into the enemy fatherland. Robert was one of 118 casualties reported from Washington that day.

From 1910 to 1920 Robert Baker's father, Wilson Van Horn Baker, known as Van, along with brother Chester were cooks at various logging camps in Skagit and Snohomish Counties including stints at Utsalady where Robert was born on November 20, 1913. In the 1920s Van Baker operated the Regis Café in Stanwood providing refreshments for the dedication of the new fire department building in 1929. In 1926, at the age of 12, Robert lost his mother Dora who died in childbirth. By 1931 Van Baker had moved to Mukilteo where he and brother Chester owned a restaurant along the waterfront. Robert Baker was a waiter in the restaurant when he married Ethel Hickey in July 1941. Eleven months later their first daughter Dora Kathleen was born.

Robert joined the global conflict late in the war enlisting in September 1943. As a private he was assigned to the 353rd Regiment of the 89th Division—dubbed the "Rolling W" after its insignia denoting its Midwestern origins during World War I. The 89th trained for six months in the swamps and backwaters of Louisiana with its first units leaving for France the day after Christmas 1944. Along the way Robert was promoted to Staff Sergeant. After more training at a camp in Normandy, the Division was trucked through France and Luxembourg and positioned to spearhead an attack deep into the heart of Germany. Mud, bitter cold, bad roads and destroyed bridges had to be overcome along with determined German resistance. The 89th, however, raced 50 miles in their first three days on the front, capturing eight villages and driving the Germans across the

Moselle River where they entrenched to make a stand. Two hours before dawn on March 16th four battalions of the 89th suffered under a hail of artillery and small arms fire as engineers placed pontoon bridges over the waterway for advancing troops. Most effective were German snipers hidden in trellised vineyards on the east bank of the Moselle. One of those is credited with ending the life of Robert Geary Baker. Ethel Baker was well into her second pregnancy during this time. A little over a month after her husband's death, she gave birth to daughter Susan.

While alive the only thing Robert Baker and Lieutenant General George S. Patton, head of the American Third Army, had in common was their duty to destroy the German Third Reich. That changed in 1945 when both became residents of the American Military Cemetery at Hamm, Luxembourg. The Staff Sergeant died protecting fellow American soldiers from the enemy—the general in a car accident in December 1945. There is little doubt that Patton would have envied the soldier's death from western Washington.

# Semper Fidelis
## Charles Isham

Charles Isham found a home in the Marine Corps. He served in the Corps for 22 years, leaving his job as a railroad clerk in Tacoma and enlisting in 1921 at the age of 17. His parents had a May/December marriage; his father being 28 years older than his mother Estella Boston Isham when he died in 1918. A year before Charles enlisted, his mother married Albert Brouillet and for the next eleven years, they would be caretakers of the hydroelectric plant on Paradise River in Mount Rainier National Park before becoming managers of the Camano City Lodge.

Soon after joining the Service, Charles was posted for several years in Peking, Shanghai and Tientsin, China serving at the American legation and on U.S. gunboats. In 1928 he was part of the American occupation force in Nicaragua fighting rebels led by Augusto Sandino, the revolutionary that inspired the Sandinista movement in the 1970s. Lieutenant Edward O'Day later reported that during a firefight, Sergeant Isham, although badly wounded, rallied his green troops with complete disregard for his own safety and while under "withering enemy fire . . . delivered volley fire at vulnerable spots in the bandit emplacements."

He returned to San Diego where he trained troops and, by the start of World War II, he rose to the rank of Master Gunnery Sergeant, the Corps' highest enlisted rank. In 1941 he became part of the 2nd Engineer Combat Battalion, 18th Marine Regiment.

Soon after, however, as 1st Lieutenant Isham, he was given temporary assignment and deployed to the South Pacific as part of a photographic support unit with the 1st Marine Aircraft Wing which snapped pictures of Japanese island positions and helped planning for American landings. The circumstances of his death on March 19, 1943 are unknown but his family was initially notified that he was missing in action—his death finally being labeled an accident. First buried on the island of Espiritu Santo and then moved to Guadalcanal, he was finally re-interred in the National Memorial Cemetery in Honolulu, Hawaii.

Charles' step-father Bert Brouillet died almost exactly a year later. He and wife Stella were associated with the Camano Beach Resort from 1936 to 1954 when she sold the lodge and restaurant.

The life of Lieutenant Isham was one of service and complete dedication to the Marine Corps where he performed various duties during his lifetime. A fitting epitaph for him may be the words given his mother in the telegram that arrived at her door: it said that her son died "to avoid giving aid to the enemy." For a Marine such as Charles Henry Isham that would have been enough.

# Onward Christian Soldier:
## The Final Mission of Gerhard Lane

*Gerhard Lane*

When Gerhard Lane left for China in August of 1935 his goal was to save souls not to take them, but history sometimes puts us on a path not of our choosing. His mission to China was the first since being ordained in the Norwegian Lutheran Church. The young 26-year-old missionary and his new wife were soon to find themselves working to survive as the world sank into the dark pit of global war.

Gerhard Almer Lane was following in the footsteps of his father, the Reverend George Lane. George Lane had been the pastor of both the Zion Lutheran Church and Our Saviour's Lutheran in Bellingham where Gerhard was born on August 11, 1909 but moved the family to Stanwood in 1918 after accepting a pastorate in that city. Gerhard appears to have been a popular student; outgoing and talented. He was elected Assistant Yell Leader of the Student Body Association of Stanwood High School in 1925, acted in the school play, was a member of the school orchestra and chosen business manager of his Sophomore class. He was also Secretary/Treasurer of the People's Luther League, a Christian youth group. Something of an athlete, in 1926 he was a starter at left forward for Stanwood's basketball team and he took first place in the city's annual bicycle race for the half- mile, winning a gold watch in 1924.

Gerhard skipped his senior year at Stanwood and instead enrolled in Pacific Lutheran College near Tacoma in 1927 where his multi-talented abilities were again evident. There he performed in the choir, worked on the school's newspaper *The Mooring Mast*, and had a lead role in the school play. His piano solo of Dvorak's "On the Holy Mount" was part of a larger program at Tacoma's Central Lutheran Church the same year. After graduating from Pacific Lutheran in 1929, he studied for a time at Luther College in Decorah, Iowa and graduated from Norwegian Lutheran Seminary in St. Paul in 1935.

Spring 1935 certainly must have felt to Gerhard as a time of great promise. Ordained as a new minister on May 27th, he married Helen Wing, a resident of Minneapolis, on June 2nd. After a honeymoon along the Washington coast, the couple set out with seven other new ministers and their families for China. Just before leaving on his mission, he gave a last sermon at his father's Phinney Ridge Lutheran Church in Seattle entitled "Persevere in Prayer." Gerhard and Helen would need plenty of both in the years ahead.

Their destination in China was the Lutheran Mission agency in Hankow which sat along the Yangtze River. The quarter-million-dollar compound had a moat-like canal at the foot of its red brick walls. Missionary Paul Frillman described trees which lined the inside bordering on playgrounds used by the students of its elementary school and the dormitories for the Chinese students— the best American school in the area, wrote Frillman.

But Japan's invasion of China in 1937 threw the area into panicked turmoil. Nanking, the country's capital at the time, quickly fell along with Shanghai, sending thousands fleeing into the interior and specifically to Hankow which became China's temporary seat of government. The battle for Hankow (today's Wuhan) would be one of the most significant of the Sino-Japanese war.

By the end of 1937, the Japanese were threatening the city. Exactly when Gerhard and Helen decided to leave Hankow for Hong Kong is not known but certainly Helen's pregnancy that summer might have been a deciding factor. Their son, John Gerhard, was born March 31, 1938 in the British colony. With his missionary work disrupted by the conflict, Gerhard Lane, by 1940, took a position as a branch manager for the International Harvester Company. With the war intensifying, however, British authorities began evacuating women and children from Hong Kong. It is likely that Gerhard and Helen, along with infant Johnny, left during that time for the American territory of the Philippines hoping to find solace in the neutral American region. They would soon discover that escape was not possible in Southeast Asia during this treacherous time.

Eighteen days before Hong Kong fell on Christmas day 1941, Pearl Harbor was bombed and the bitter winds of war once again found the Lane family. The Japanese invasion of the Philippines began ten hours after the attack on the Hawaiian base. On December 22nd the Japanese landed on the northern area of Luzon pushing the Americans south in desperate fighting. At that point, Gerhard Lane offered his services to the army. He was initially given a commission as a first lieutenant and assigned to the headquarters company of the 81st Infantry but was quickly promoted. Gerhard, in his last letter home, wrote his mother: "We've all had to work hard during these first few months since the war struck. It's a

44

hard, bitter job, but it must be done. Our job now is to wipe out the Japanese—and every white man, Chinese and Filipino over here must help!"

The Japanese broke the American lines in April 1942 and invaded the Bataan Peninsula on May 4, 1942, landed May 5th on the island fortress of Corregidor in Manilla Bay and occupied the American stronghold on May 7th forcing General Jonathan Wainwright to surrender all American troops throughout the archipelago. Brigadier General Guy Fort had at the start of the year taken control of the 81st Division of the Philippine Army (jointly American and Filipino) on the country's southernmost island of Mindanao where he also organized Moro tribesman to resist the Japanese. Captain Gerhard Lane was an integral part of that effort.

Like General Fort, teacher Edward Kuder was an expert on the Islamic Moro people of Mindanao having worked with them for over 20 years. The Moro were highly independent, incredibly brave and fierce fighters who took naturally to being guerrillas, said Kuder. They became the core of resistance to the Japanese invaders on Mindanao inflicting, in some cases, significant casualties on the enemy. However, there were many, said Kuder, who would readily turn outlaw. Sometimes anti-American or simply bad actors, they used the chaos of war to their advantage, committing outrages such as attacking Christian Filipinos as well as deceiving or betraying unwitting Americans.

Assisted by a Moro headman, Kuder was in hiding after the fall of the Lanao provincial capital Dansalan, on May 2nd. He hoped to join General Fort who had established a camp in a heavily forested area after his army had splintered into guerrilla units following initial fighting with the Japanese. It was Lane, however, who brought Kuder orders from General Fort to stay put and coordinate Moro assistance in securing supplies for guerrillas. Gerhard had been instrumental in persuading Chinese merchants to turn over hidden stocks of supplies of all kinds just before the Japanese landings. Escaping advancing Japanese troops, American and British civilians, including Helen and Johnny Lane, were hurried out of Dansalan under the care of a Catholic priest assisted by U.S. Naval Commander Walter Bicknell. Moving to a momentary area of refuge, Bicknell later reported they were accosted by outlaw Moros who, while acting as cargo bearers, ran off with much of their supplies.

Fort finally received word that Major General William Sharp had ordered all American forces on Mindanao to surrender. Although strongly disagreeing with the order, Fort complied. Bicknell, now accompanied by Captain Lane, began the trek back to Dansalan with the civilian refugees where they would surrender to Japanese authorities. On the evening of May 27, 1942, they halted for the night at a Moro village along the shores of Lake Lanao. Captain Lane placed guards around the camp perimeter. During the evening sounds of a scuffle attracted his attention and while investigating he was stabbed to death by a Moro friendly to the Japanese who then made his escape. His body was buried in the Catholic cemetery in Dansalan.

Helen and Johnny Lane spent the rest of the war in Japanese prison camps on Mindanao. After the war Helen became a music teacher in Hawaii until her death in 1990. In 1949 the body of Reverend and Captain Gerhard Lane was disinterred and reburied in Pacific Lutheran Cemetery in Seattle against the expressed wishes of his angry wife. He was posthumously awarded the Purple Heart but his field promotions to captain and later to brevet major were overlooked in his file and never officially approved by the Army.

Johnny Lane moved to Florida later in life and died there in 2012. He still had the Purple Heart of the father he barely knew. That medal is now back in Washington state having been purchased by a collector of militaria from a neighbor of John Lane. Christian soldier Gerhard Almer Lane was thrust into a world he had not imagined. There was more of the minister in Gerhard's death than army officer—helping people had been his cause. But no one could say that he did not fulfill his duty for both God and country.

# A Short but Devout Life
## Roger William Barney

It was family tragedy which brought Roger Barney to the
Stanwood area. His mother Eleanor Neslund Barney died in 1928
while the family lived in Everett. His French-Canadian father
William had left the mills of Everett by 1926 to become a police
officer in the city. Roger and his sisters are listed as living in an
Everett boarding house in 1930 although their father is listed as
living alone in the census. William Barney remarried in 1931 and
it appears that his children were farmed out to extended family
members. Roger would be taken in by his Uncle Cliff and Aunt
Elizabeth Neslund on their dairy farm in the Cedarhome area of
East Stanwood. Clifford's Swedish mother, Roger's grandmother,
Christina also helped in raising the 12-year-old boy.

Roger attended Lincoln High School where he was a star athlete
in basketball and softball; and "because of his wholesome way
of living he excelled in all of them," as reported by Stanwood's
*Twin City News*. He became a devout member of the Cedarhome
Baptist Church attending services at home and while away. The
newspaper reported that Roger "had a deep and unshakeable faith
in God and found much comfort and strength in the reading of the
Word and prayer throughout his Christian life." He reportedly
was as talented in his studies as he was athletically.

However, after one year at Mount Vernon Junior College, Roger
enlisted in the Army Air Corps in October 1941, three months
before the attack on Pearl Harbor. He was assigned to the medical
corps and sent to Camp Grant at Rockford, Illinois for training.
Following the U. S. entrance into World War II, Roger was
assigned to Stockton Field in Stockton, California but soon became
ill and was sent to Hammond General Hospital at Modesto.

The news reported that Roger "was sincerely grateful for the
many letters and gifts of flowers sent him during his illness by
friends and loved ones, but was unable to acknowledge them

because of the seriousness of his illness. He was especially appreciative of the many who wrote that they were praying for him, which he said helped him more than anything else." Penicillin and numerous blood transfusions failed to improve his condition and the young soldier died a little past 8 a.m. on June 1, 1944 if hemolytic anemia.

# "Death Was No Stranger"
## Harold McCann

Harold McCann did not grow up in Stanwood. However, he was born there on August 28, 1920 while his father John worked as a laborer for the Carnation Milk condensery in East Stanwood. John McCann was from Everett by way of Michigan where as a boy John's father ran a boarding house. John married Sigrid Helland in July 1919 and they lived with Sigrid's widowed mother Bolleta who ran a small dairy operation along the Stillaguamish River. By 1930 the McCann family was back in Everett where John and Sigrid ran a grocery.

Harold graduated from Everett High School and joined the Army in March 1940 but quickly is shown as a patient at the Fort McDowell Military Base in Sausalito, California. Initially assigned to a coastal artillery unit, he later became part of the Philippine Department's Headquarters Company—an assignment that would become a nightmare. With the invasion of the Philippines in December 1941, Harold and thousands of other troops fought desperately against a ruthless enemy. Bataan fell April 9, 1942 and the island fortress of Corregidor May 6, 1942. Harold was reported as severely injured four days earlier.

Roughly 11,000 captured Allied troops, many wounded, were force-marched to Manila's Old Bilibad Prison where food and water were scarce or nonexistent. They were then moved to Camp

Cabanatuan in May, where American dead in the camp were left to decay in the intense heat as exhausted prisoners lacked the strength to bury their own. An estimated 30 Americans died each day at Cabanatuan.

In October of 1942 about 1,000 prisoners were put on cramped freighters and shipped to the Davao Prison Colony on Mindinao Island. It is likely that Harold McCann was in this group since later reports cite Davao as the place of his death. His parents were notified in January 1943 that he was a prisoner of war. The War Department officially reported Private McCann dead on July 9, 1943—one of thousands of prisoners who succumbed to conditions in the death camps.

Accounts by survivors point to the abject cruelty of the Japanese captors forcing them at bayonet point on a 17-hour march from the dock to the Davao Prison. Many never finished the march. They were told that their easy life was over. At Davao, they were forced to labor in the fields and logging mills under treacherous conditions. Medicines were virtually non-existent and prisoners deteriorated rapidly perishing by a variety of diseases including beriberi, malaria and dysentery as well as general malnutrition or abuse. The only large-scale escape from a Japanese prisoner camp in World War II, was from Davao on April 4, 1943 and then only by three men. Smuggled to American authorities by Filipino guerillas, these prisoners were the first to report the inhuman conditions—the "heart-breaking truth,"—of the horror inside the camps.

Of over 36,000 American military personal taken as prisoners in the Pacific Theater, over 13,000 died as POWs. Navy Commander Melvyn McCoy later wrote that "the hardships we had faced in battle were, if anything, much less severe than those awaiting us as military prisoners...Death was no stranger to any of us who had gone through the Battle of the Philippines."

# Tablets of the Missing
## Donald Garrison

*Donald Garrison*

Some lie in deep, watery graves, some under the canopy and debris of dense jungles and others in burial sites left unmarked by captors. They are all America's MIAs. For those Missing in Action during World War II in the Pacific, their names are etched on gray, stone walls at the Manila American Cemetery in the Philippines—called the Tablets of the Missing. One name listed is 2nd Lieutenant Donald Howard Garrison. The naval aviator gave his life on September 22, 1943—only his second mission of the war.

Lieutenant Garrison's family had deep roots in Virginia but a long history with Camano Island and Skagit County. His great-uncles David, Tom, Andy and James Esary began as loggers in Seattle and later pursued farming and business interests after arriving in the area in 1883. In 1898 they began a large logging enterprise around Camano City. Their 17-year-old nephew Porter Garrison, Donald's father, joined them in 1900. Initially a cook for the camp, Garrison would manage the site by 1920. In later years Porter Garrison and his sons were instrumental in the logging and real estate business on the island. Their logging business took them to areas throughout the northwest including British Columbia.

Donald was born in 1922 in Everett but graduated high school in Victoria, British Columbia before attending the University of Washington where he played football. He was 19 when he enlisted in the Army Air Corps in April 1942 as seemingly unstoppable Japanese forces spread their tentacles of conquest across Asia and the South Pacific. His basic and primary training was at the Army Air Base in Santa Ana, California beginning in August 1942. His letters home show his excitement and optimism but also the challenges to pass the necessary requirements and become a naval fighter pilot. "If I wash out any where a long the way it will be OK. Because I'll know that I tried and just didn't have the ability," he wrote his family. Those who did flunk out of this school usually went on to assignments as bombardiers or navigators. Donald, however, wanted to be a pilot. He hoped to fly a P-47 fighter plane but admitted he had never been in one although he loved their look.

Training and the mundane chores of army life were punctuated by weekend leaves. He visited Hollywood and the footprints of the stars but fell behind in his studies after being hospitalized. He had complained of illness, but vowed to know better next time. Success at Santa Ana led to his transfer in January 1943 to the Polaris Air Academy near Lancaster, California in the Mojave Desert, used by the government for new American and British pilots. Instructors mainly used a Vultee BT-13 for training. Donald Garrison was finally in the air practicing formation flying

*Tablets of the Missing*

and landing. His college sweetheart and fiancée Madge Niccum came to visit but had to stay in Ontario, California, an hour and a half south of the base, "so I guess I won't see much of her." His final advanced training would be at Williams Air Force Base near Chandler, Arizona. There on May 1, 1943 he and Madge were married. Their only child, Donald, Jr. was born in April 1944.

It would not be P-47s for Donald but the Lockheed P-38 Lightning. The aircraft with its greater range and dual engines was used most extensively in the Pacific Theater especially for long missions over water. Following American naval victories at Coral Sea and Midway, American forces drove the Japanese from Guadalcanal and began the arduous and bloody campaign to liberate New Guinea—just north of Australia—the second largest island in the world. The Japanese were capable of large-scale attacks sending over 100 aircraft against the Allies at Port Moresby in April 1943. Donald's Fighter Squadron, the 432nd, which flew exclusively P-38s, was activated in May of that year. They were the only planes with enough range to hit Japanese bases on the island's north end. He arrived at Port Moresby in August 1943.

The young pilot had little time to get comfortable, flying his first mission on September 20, 1943 as the wing man for Lt. Thomas Simms. Returning from flying escort for B-24 Liberator bombers, they went to the aid of fellow P-38s engaged in a dogfight with enemy aircraft. "We were flying back pretty much alone, not close formation, with my wing man Lt. Garrison. We got to about Madang on the coast of New Guinea. Some of the P-38s below us were getting hit by the Japanese, asking for some help, so we made a 180 and saw some enemy airplanes way off, up at our altitude, doing 'yo-yos.' I peeled off … to give the guys below some help." Simms' plane was hit from behind by an enemy fighter. "I felt the impact of his bullets it felt like people with hammers hitting the armor plate behind my head!" Simms was shot down but survived and was later secreted back to Australian forces by a local native.

Donald Garrison made it back to his base at North Borio but two days later took to the skies in support of an amphibious landing at Finschafen. The subsequent air battle cost the Japanese seven bombers and eleven enemy fighters. Only two P-38s were lost that day and only one pilot—Donald Howard Garrison. Captain Daniel Roberts later remembered that while pitched in a steep dive from 20,000 feet, he saw Garrison's right engine smoking badly. Lt. Garrison did not respond to radio calls by Captain Roberts. The young lieutenant was never seen again. Madge Garrison received the dreaded telegram on September 27th: "I regret to inform you …" The telegram was followed by a letter from the unit's commander, Major Frank Tomkins, on October 8, 1943. Although offering a small glimmer of hope that Donald might have survived a crash in the jungle, his sentiments revealed the truth. "Don had been with us only a short time, but in that time we had learned to love and admire him for his good character, friendliness and his willingness to meet the enemy. His loss to the squadron is felt very deeply by both his fellow officers and enlisted men."

Donald's mother wrote him five days before his final mission. We don't know if he ever got to read it. She relayed stories from home of family and friends and was happy he had arrived "somewhere" safely. She promised that she would try not to worry but kept him "much in mind and pray for you as you go forth to the task assigned you—that you may come home to us fine--noble, safe and well—and an honor to your country and home." Donald Garrison was posthumously awarded the Air Medal and Purple Heart. He was officially declared dead on December, 19, 1945. In 1994 his brother Bill and a niece spent a week in New Guinea searching unsuccessfully for any information on the wreckage of Donald's P-38.

# A Self Made Man
## Leonard Broin

*Headstone of Leonard Broin*

Len Broin's 1929 Stanwood High School yearbook called him a "self-made man" whose ambition was to have a money tree. World War II interrupted that whimsical goal. Leonard Martin Broin was born in Fertile, Minnesota in February 1912 and moved with his family to East Stanwood in 1916. Len enlisted in the U.S. Navy September 30, 1942 and joined the crew of the minesweeper YMS-102 December 5th of that year. YMS-102 was completed by the Astoria Marine Construction Company on August 22, 1942. The wood-hulled YMS class ships were used to sweep for underwater mines as well as potential hazards along American shorelines or overseas naval bases. The "Y," referring to Naval Yard, was a designation that distinguished these ships from other minesweepers as they generally worked waters adjacent to naval facilities.

One of his shipmates was Frank Curre, Jr. Curre was a survivor of Pearl Harbor aboard the battleship, U.S.S. *Tennessee*. Re-assigned to the YMS-102, Curre thought there must have been a mistake. "When it's time to check in to the new ship, I go down there to the chief petty officer on duty, and I showed him my orders. I said, 'Where is this ship, Chief?' He said, 'Go down to the end of the pier. It's right down there.' Well, YMS-102 is on the bow of this thing. It stood for yard minesweeper. I didn't go far enough to see that, so I go back and tell the chief. I never got over that. I said, 'Chief, you're mistaken. There ain't nothing down there but a fishing boat.' He come back to me and said, 'I tell you what you do, son. You go get all your fishing gear together and you come back because that's the one you're going on.' Man, that thing wasn't much bigger than our motor launches on that battle ship. I stayed on there '42, '43."

An anonymous sailor left a poem to describe his time aboard a YMS:

**A Plug for a Distinguished Nervuos [sic] Cross**
Listen, men, I've a tale to tell,
of mighty midgets that sail like - well,
with a word to the wise on larger ships,
to forget those small craft transfer slips,
Men don't live on YMS's -
they just exist under strains and stresses,
tossed around like a bundle of peas,
inside their ship on the calmest seas,
Did you ever eat on a YMS?
It has been done at times I guess,
but the simplest meals can come to grief,

when we hit the wake of a floating leaf.
An order comes to dog the hatches,
for days on end we all wear patches,
what dire calamity caused all this?
A passing school of playful fish.
Then, at "0 two hundred" all's secure,
the anchor is deep and sure,
and even when the seas like granite,
she's taking off for another planet.
The battered life is just one item,
we've many more, just let me cite 'em,
We scrub our whites - they come back black,
our clothes line boys is aft of the stack.
The spacious lockers, I might mention,
are always full and gosh, the tension.
I wish the Navy were more lenient,
four rubber sides would have been convenient.
I'm not through with this little tale,
of little ships and how they sail,
half submarine and aeroplane,
they're a secret weapon gone insane.
Ah yes, my friend, if big ships bore you,
the YMS is waiting for you,
with loving care, from fore to aft,
the Navy designed them and laughed and laughed.
Author Anonymous


(Courtesy of Robert Noonan - YMS-176, U.S. Navy Minesweeper - World War II - Pacific Area)

Curre was Broin's shipmate for the short time Len served. Both joined the ship in Seattle which sailed December 28, 1942. It operated for a time out of the Naval Air Station on Whidbey Island, Washington before being transferred to Pearl Harbor, Hawaii. Len was promoted to Boatswain Mate August 1, 1943. He became ill and was transferred to the naval hospital on October 12, 1943 and died October 30th. He was returned to the U.S. and buried in the Golden Gate National Cemetery in 1947.

# Death Along the Rapido
Peter Rekdal

*Peter Rekdal*

Before the Allied assault across Italy's Rapido River in late January 1944, Major General Frederick Walker stated that such "a frontal assault across the Rapido would end in disaster. . . . I am prepared for defeat." Walker's 36th Division of the American 5th Army would spearhead the attack. The failure became known as one of the greatest tragedies suffered by the U.S. Army in World War II. Of the roughly 4,000 men in the 36th, mainly Texas National Guardsmen, roughly half were either killed, wounded or missing at the end of the bloody three day attack. Many blamed the near suicidal assault on the 5th Army's commander, Lieutenant General Mark Clark. The 19th Combat Engineer Regiment, which supported the Texans, also suffered heavy casualties. One of those from the 19th killed during the battle was a young man from Camano Island, Peter Trygve Rekdal. Peter had just turned 33 years of age on January 4, 1944.

Doubtful Peter had ever heard of the Rapido River or nearby Monte Cassino prior to his landing in Italy. A year before America's entrance into the war, Peter was working for the Saginaw Logging Camp near Prosser, Washington. Ten years before he lived with his parents, Trond and Anne, on Livingston Bay. His May 1942 enlistment lists skills as a lumberman and raftsman. He had seen tough fighting before Italy, surviving the invasion of North Africa and Sicily. At Kasserine Pass in Tunisia, where Americans suffered an initial devastating defeat, Peter's regiment was in the heart of the battle. In a May 1943 letter home Peter wrote that "they [the Germans] gave us quite a shaking up, but got the worse of it in the end, anyway." Peter's other requests were common among soldiers in the field including sending money for a sleeping bag—"if one can be found that's light enough to pass shipping regulations. I certainly don't need one right now, but imagine that by the time it reaches here it will be leaning toward winter again. Had a fairly good one through most of last winter but was forced to donate it to some Jerry at Kasserine. I believe that Clyde Tolin at Hartney's could get hold of one if there are any to be had, so just let him take the trouble off your hands." The 19th Combat Engineers had 128 casualties, seriously reducing its effectiveness.

The crossing of the Rapido on the night of January 20th was meant to tie down German troops and prevent their use against an Allied amphibious landing at Anzio two days later. The risky strategy was chosen because defensive fortifications of the German Gustav Line had thwarted the Allied advance up the Italian peninsula. The terrain, including swollen rivers, ever-present mud and marshes favored the Germans. The fast flowing, icy waters of the Rapido would be difficult enough, but the Germans had also flooded many areas and cleared a mile-long killing field leading to the river, made ever more deadly by stretches of mine fields. The nearby garrisoned Monte Cassino, a historic mountain abbey, gave the Germans high ground from which to direct their artillery.

Units of Rekdal's combat engineers were assigned to individual regiments of the 36th. Their job was to build foot bridges at the river's edge and carry boats to be used by the attacking Americans. The intense shot and shell of German gunners produced a landscape of carnage. Sergeant Billy Kirby later recalled that he "had never seen so many bodies of our own guys. Just about everybody was hit. I didn't have a single good friend who wasn't killed or wounded."

Peter died early in the struggle on January 21st. His commanding officer, Edgar Pohlmann, wrote his parents. Peter performed with courage, Captain Pohlmann wrote; "he died heroically." Pohlmann said that Peter was a friend to every officer and enlisted man. "Words in themselves cannot in any way express my profound sympathy to you, the bereaved father but in a small way," the captain continued. "I sincerely hope this letter will in some measure assuage and alleviate your poignant suffering and great loss." Peter's memorial service at Our Saviour's Lutheran Church in Stanwood was described as "impressive."

# Hell Ship Horror
## Ernest Moser

The 1937 Stanwood High School yearbook acclaimed Ernest C. Moser as a "football hero." Not bad for a young man who only took up residence in the town a year earlier. The seemingly light hearted student professed that "when work interferes with play— away with work." Born in St. Helens, Oregon on April 1, 1919, he spent most of his early years bouncing between Washington and California where his father took jobs as a freight driver and lumberman.

Perhaps because of his nomadic upbringing, Ernest soon enlisted in the Army Air Corps after graduating in 1938. On November 14, 1940 he joined the service at San Pedro, California and was stationed with the 30th Bombardment Squadron, 19th Bombardment Group Heavy, in the Philippine Islands. His timing could not have been worse. Within hours after the disaster at Pearl Harbor, Japanese troops invaded the Philippines on December 8, 1941. The last American stronghold of Bataan on Manilla bay surrendered April 9, 1942 and Ernest's real nightmare began when he became a prisoner of the Japanese—his parents notified in May 1942. Their last notification reportedly came on

*Ernest Moser*

the day Ernest died, September 7, 1944. A year earlier they had received a censored post card from the prison camp where Ernest languished. The card spoke of his good treatment. "Don't worry," it said, "I'm O.K....not wounded."

It is believed that he ended up in the camp near Cabanatuan, Luzon. Conditions there, as in other Japanese camps, did not mirror the positive spin of Ernest's post card insisted upon by his captors. Fellow prisoner Vice-Admiral Ken Wheeler remembered the filth, poor food, disease and brutality which plagued those incarcerated. "The majority of the prisoners had malaria or dysentery or both, and medical care was virtually hopeless since our own doctors were sick as well and none had enough medicine to really help....Our pleas for help went completely unheeded.... We lost upwards of thirty men each day for the first three months, not counting those who were executed," he said.

In early 1944 American forces intensified bombing of the Philippines in preparation for a landing later that year. Many of the prisoners at Cabanatuan were transferred to Mindanao and loaded onto old merchant ships for transport to Japan to be used as slave laborers. Ernest Moser was put aboard the antiquated *Shinyo Maru* along with 750 other Americans. These decrepit floating prisons were dubbed "Hell Ships" by the men who suffered on them. The POWs were placed in the dank, foul depths of a lower compartment with a small ladder as the only exit. Word quickly circulated that the Japanese vowed to kill the prisoners if the ship was attacked.

Those fears were realized on September 7, 1944 when two torpedoes slammed in quick succession into the ship. The commander of the submarine USS *Paddle* had been incorrectly informed by military intelligence that the ship carried Japanese troops. First Lieutenant John J. Morrett remembered struggling through the mangled bodies of fellow prisoners after the explosion. Japanese troops opened fire with machine guns on those men making their way onto the upper deck. Others were shot after jumping into the sea. Only 82 Americans survived the ordeal, making it to nearby islands or were picked up by the *Paddle* following the sinking of the *Shinyo Maru*. Ernest Christian Moser was not among them. His body was not recovered but his name is etched on a monument, along with other deceased servicemen, at Manila's Fort William McKinley. For his service of 335 days as a member of the Air Corps and his 854 days as a prisoner of war, Ernest was awarded the Purple Heart.

After his war's end, his mother Mae Moser received a letter from the commander of allied forces in the Pacific:

My dear Mrs. Moser:

My deepest sympathy goes to you in the death of your son, Pvt. Ernest Moser. You may have some consolation in the memory that he, along with his comrades in arms on Bataan and Corregidor and in prison camps, gave his life for his country. It was largely their magnificent courage and sacrifice which stopped the enemy in the Philippines and the final defeat of Japan. Their names will be enshrined in our country's glory for ever.

In your son's death I have lost a gallant comrade and mourn with you.

Very faithfully,

Douglas MacArthur

# The Deadly Blossoms
## Gordon Lord

The temperature inside an unpressurized B-24 Liberator bomber flying at 25,000 feet can often drop well below zero. Layered in flight suits and flak vests, crews required rubber oxygen masks fitted to their mouths and noses above 10,000 feet. Tough work environment if you're a waist gunner on such an aircraft. On September 13, 1944, that job belonged to a native son of Stanwood, Sergeant Gordon Leith Lord, aboard the bomber *Ain't Bluffin'*. Mix in attacks by German Messerschmitt fighters and the deadly black blossoms of anti-aircraft flak that peppered the skies above Germany with lethal blasts of shrapnel, and job longevity was at a minimum.

The 22-year-old Lord was a member of the 853rd Bomb Squad, 491st Bomb Group stationed in North Pickenham, England; part of the 8th Air Force. Roughly half of all the Army Air Corps' 47,000 casualties, including 26,000 deaths, were suffered by the 8th Air Force tasked with destroying the infrastructure of the Third Reich. The 491st flew 187 missions and lost 47 bombers including *Ain't Bluffin.'*

*Gordon Lord*

Gordon Lord was born January 15, 1922 in Stanwood to Anna K. Pearson and William Leith Wade. Anna's parents Albert Pearson and Marie Hansen Pearson farmed in Cedarhome. Anna and William Wade divorced in 1924 and Anna married Willard Lord who three years later moved the family to Wapato in Chelan County. In 1940 Willard was service manager for an electric company and both Anna and her son worked in the area's apple orchards. Two months after enlisting in the Army Air Corps Gordon married Betty Loraine Halstead in April 1943. A year later the 491st Bomb Group shipped out for Europe.

Their mission on September 13, 1944 included strikes by 342 B-24s on the Schwabish Hall Airfield and munitions dumps at Ulm and Weissenhorn. At speeds of roughly 300 miles per hour, the bombers flew in staggered tactical formation creating a "combat box" of interlocking fields of defensive fire for combatants, such as Gordon, who hunched over mounted 50 caliber machine guns.

While fighters were a danger, by late 1944 Germans depended more on a massive antiaircraft system which was the greatest threat to U.S. bombers on the 13th of September. A direct hit on *Ain't Bluffin* exploded its number two engine thrusting the plane sharply down and to its left. Its left wing pierced the fuselage of another aircraft, *Time's A-Wastin'*, ripping away most of that plane's tail section and pushing the aircraft into a nosedive. Moments later *Time's A-Wastin'* burst into flames, killing all but two of its crew. Six crew members aboard *Ain't Bluffin'* bailed out of the plummeting aircraft and were taken prisoner. Three others including Sergeant Lord went down with the aircraft crashing near Sermenstedt, Germany.

Only one in three airmen survived the air battles over Europe in World War II. Their plight was the subject of World War II vet and American poet and novelist Randall Jarrell:

### Losses
It was not dying: everybody died.
It was not dying: we had died before
In the routine crashes-- and our fields
Called up the papers, wrote home to our folks,
And the rates rose, all because of us.
We died on the wrong page of the almanac,
Scattered on mountains fifty miles away;
Diving on haystacks, fighting with a friend,
We blazed up on the lines we never saw.
We died like aunts or pets or foreigners.
(When we left high school nothing else had died
For us to figure we had died like.)

In our new planes, with our new crews, we bombed
The ranges by the desert or the shore,
Fired at towed targets, waited for our scores--
And turned into replacements and woke up
One morning, over England, operational.

It wasn't different: but if we died
It was not an accident but a mistake
(But an easy one for anyone to make.)
We read our mail and counted up our missions--
In bombers named for girls, we burned
The cities we had learned about in school--
Till our lives wore out; our bodies lay among
The people we had killed and never seen.
When we lasted long enough they gave us medals;
When we died they said, 'Our casualties were low.'
They said, 'Here are the maps'; we burned the cities.

It was not dying --no, not ever dying;
But the night I died I dreamed that I was dead,
And the cities said to me: 'Why are you dying?

We are satisfied, if you are; but why did I die?'

# Extraordinary Valor
## Robert Nelson

Robert Kolden Nelson and his grandfather Nels Olsen died within eight months of each other in 1944. Robert was born in Everett and grew up in Bellingham. His grandfather Nels, however, had deep roots in Stanwood where he lived for over 50 years.

Nels Olsen emigrated to the U.S. in 1886 from Norway at the age of 18. He married Petra Kolden in Seattle in 1895 and they moved to Stanwood where Nels became a long time employee of the Stanwood Lumber Company rising to the position of foreman. A city councilman under Mayor D. O. Pearson, he served as the town's mayor in 1921-1922, and was a Justice of the Peace and Police Judge through much of the 1920s and 1930s. As a judge

*Robert Nelson*

he was known for his impartiality and "fearless honesty" in administering the law. Nels' daughter Anna married Roy Nelson in July of 1921. Roy was a Navy veteran of World War I and worked as a driver for the North Coast Transportation Company. The couple lived in Bellingham.

Robert was commissioned while attending Western Washington University. Twenty-one-year-old Ensign Robert Kolden Nelson was assigned to the U.S.S. *Enterprise* as a fighter pilot having won his wings at the age of 19 following training in Corpus Christi, Texas. During the landings of American forces in the Philippines, Nelson flew air support for troops. He earned his status as an ace after downing a Japanese fighter plane over Manila on October

18, 1944. By the time of his death nearly a month later he was credited with six confirmed kills of enemy aircraft and assisting in downing four others. His Squadron (VF-20) was the only one to engage all three of the Japanese attacking forces during the historic Battle of Leyte Gulf. He was awarded the Navy Cross, that Service's highest award, for his attack and damage of an enemy light cruiser on October 25, 1944. His citation read that he made a direct hit on the cruiser while plunging through a "fierce hail of anti-aircraft fire" which riddled his engine forcing him into the sea. That time he was rescued.

He was not so lucky 25 days later. On November 19th four Japanese attack bombers, nicknamed Bettys, closed in on American Task Force 58. American aircraft on the *Enterprise* scrambled to intercept them including Robert Nelson in his Grumman F6F Hellcat fighter. Nelson started a run on one of the Bettys taking fire from its tail and top gunners. His left wing in flames, Nelson pulled up steeply before his aircraft rolled over and dove into the sea. His body was never recovered. The Japanese attackers were all destroyed before reaching their target.

The Battle of Leyte Gulf is believed to be the largest naval battle in American history. Robert Nelson's meritorious performance as a Navy pilot earned him ten medals altogether which included the Distinguished Flying Cross for extraordinary achievement under fire.

# A Cottage at Warm Beach
## Donald Leach

The Leach family was well-known in Arlington. Its patriarch, English born E. Clement Leach, was a respected physician and community leader who had immigrated to the U.S. in 1900, receiving his medical training in Pennsylvania. He took over the

Donald Leach

Arlington General Hospital in 1918 which the local newspaper called "almost a sanctuary for the ailing and injured." Special interest was given to children "holding that when he serves the children he really does something worth while for the people among whom he lives." Dr. Leach was a member of the Arlington school board for 20 years.

His oldest son Donald was born in 1915 in Connecticut prior to the family moving to the West Coast. The family including brother Robert, sister Marjorie, mother Irma and their shaggy dog Abby, were fond of spending summer days at their Warm Beach cottage—"summer people" as one author called them. By all accounts, Don Leach was a well-adjusted and bright student. He was in school plays at Arlington and competed for a county scholarship in Chemistry before entering the University of Washington where he pledged the Kappa Sigma fraternity. Along with family he journeyed to California in 1937 to root for the Huskies in the Rose Bowl.

Don Leach did not continue his schoolwork, however, dropping out of school after one year. In 1939 he married Louise Speer of Bellevue and together they made their home in the little summer cottage on the shore of Port Susan. Don is listed as a butcher in a local slaughtering house in the 1940 census. In December 1943 he was drafted and joined the 2nd Infantry, part of the Army's 5th Division.

The 5th Division was a spear point in the fall of 1944 as George Patton's Third Army attacked the city of Metz, a gateway to the 22 forts of Germany's Siegfried Line. The city lay between two rivers and itself was protected by several forts and observation points connected by entrenchments and tunnels. During the 11-day drive, the 2nd Infantry lost 48 men killed in action, five officers and 286 wounded. Don Leach is believed to have fallen to enemy gunfire on December 7th, three years to the day since the beginning of the war. The last of the city's fortifications surrendered six days later. Records often list Don Leach's hometown as East Stanwood—a probable nod to the little Warm Beach cottage that held both Don's childhood memories as well as those of a new marriage.

A 1935 news article praised the selflessness of Dr. Clement Leach—"When our loved ones go down into the shadow of death, we turn to these physicians of ours with hope and confidence." They are "well skilled, kind, faithful and true to our service." It went on to say that "as men, we owe them much—more than money will ever repay." Those words hold true not only for doctors but for our veterans as well.

# Two Gold Stars
## Dorman Riker and Andrew Riker

Millie Pigort received a Purple Heart in early 1945, posthumously earned by her son Andrew Riker, known as Bud, who died in a French field on December 10, 1944. The news was hard on Bud's twin sister Mollie who remembered Bud as her best friend.

Bud Riker was a private with the 17th Armored Infantry Battalion, part of the 12th Armored Division, which breached German defenses at the Maginot Line near Rohrbach, France where he fell. He graduated from Lincoln High School in East Stanwood in 1942. He rests today in France's Lorraine American Cemetery.

The family would soon learn the fate of another Riker brother and Lincoln grad, Dorman. He had been taken prisoner after Japanese

*Andrew "Bud" Riker*

forces seized Manila on May 6, 1942. He joined the Navy in 1938 and attended the Naval Academy a year later. A Pharmacist's Mate 2nd Class, he was then sent to Quantico, Virginia and trained as an X-ray technician and there became engaged to Mary Landrum of Philadelphia.

After his capture Dorman remained as part of the medical staff at the Canacao Naval Hospital in Manila, part of the Bilibid Prison compound. Personnel treated the endless number of prisoners suffering from everything from amoebic dysentery and tuberculosis to beatings and gunshot wounds inflicted while part of Japanese labor gangs. As U.S. naval planes began bombarding the Manila area in September 1944, the Japanese began a draft of prisoners to be sent to Japan. Dorman was put on the "Hellship" *Arisan Maru*—one of a fleet of aging freighters used for transport. Over 1,700 men went aboard on October 10, 1944. They were crammed in the front holds of the ship with barely enough room to move or lay down. Most had to squat or stand during the next two weeks at sea, existing in near total blackness, stagnant heat and the fetid stench of human filth. Many were ill, some died--some went mad. Many prayed to die. On October 24th, the U. S. submarine *Shark* attacked the convoy in the South China Sea which included the *Arisan Maru*, not realizing that it carried Allied prisoners. Three torpedoes were loosed but only one hit the freighter amidships on its starboard side. Rope ladders to the prisoners were cut by the Japanese who abandoned the ship. Some prisoners did escape and lowered ropes to the others. Those strong enough went into the chilly waters as the ship began to sink, swimming toward Japanese destroyers and lifeboats only to be beaten back by their captors. Nine men would survive to tell the horrific story of the *Arisan Maru*. It is America's worst naval disaster in terms of lives lost.

*Dorman Riker*

Researcher and writer William Bowen's father also died that day on the *Arisan Maru*. He found the inscription on Corregidor's memorial a proper benediction:

"Sleep, my sons, your duty done, for Freedom's light has come; sleep in the silent depths of the sea, or in your bed of hallowed sod, until you hear at dawn the low, clear reveille of God."

# Held in High Esteem
## Edward Pearson

Ed Pearson was likely very sick even before he disembarked at Southampton, England October 22, 1944. He was with the 309th

*Troop ship, SS. John Ericsson, Library of Congress, LC-DIG-ppmsca-19287*

Infantry Regiment, 78th Infantry Division destined to be thrown into the chaotic scramble of Hitler's last push to save the Reich dubbed the Battle of the Bulge. They had boarded the troop transport SS *John Ericsson* in New York nine days earlier. The *Ericsson* was a converted Swedish passenger liner formerly called the *Kungsholm*. These ships normally would hold a bit over 2,000 passengers but the exigencies of war put 12 to 15,000 men onto the ships in the rush to supply American troops in Europe.

Ed Pearson's death was an anomaly at that time. Medical officials cite two eras for the deaths of American servicemen and women. During the "Disease Era," from 1775 until 1918 more soldiers died of diseases than were killed on the battlefield. A ratio of five to one during the Spanish American War and two to one during the Civil War. The heightened threat of the 1918 pandemic resulted in taking half of all the U.S. casualties in Europe during World

War I. Things had changed dramatically for the better by World War II. The advent of antibiotics, improved medical training and resources changed much. The numbers for the "Trauma Era," which continues today, shows the predomination of combat related deaths over disease.

For Private Pearson, however, his war was to end in a field hospital on December 13, 1944. His death was from complications following a hernia operation that occurred soon after being drafted on December 28, 1943 after turning 18. Hospital admission records lack any specificity of his death: inflammation of the kidneys (nephritis); purpura (internal bleeding) or possibly pneumonia. It made little difference for the young man from East Stanwood, Washington.

Edward Henry Pearson was born April 22, 1925 to Swedish parents in Battleview, North Dakota where his father Jöns farmed. By 1935 the family had re-located to an area dubbed the Village in East Stanwood, Washington. At that time Ed's mother Thilda appears to be supporting the family by her work for the Works Progress Administration on road projects. Later they would move farther south near the community of Silvana and the nearby Stave's Garage.

Ed attended Lincoln High School in East Stanwood and found work after graduation in the shipyards of Stanwood along the Stillaguamish River. The yards produced barges for the U. S. Army until early in 1945.

Pearson's obituary pointed to the loss felt by his classmates at Lincoln saying that he "will be grieved by the boys and girls with whom he attended school, his pleasing personality and splendid character winning the esteem of all."

# "First In, Last Out"
## Floyd Perin

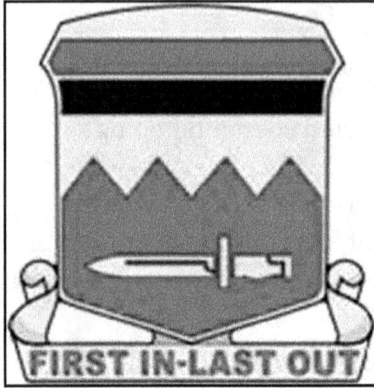

Floyd Perin Jr. may not have remembered his father well but the trauma of his father's death may have stuck with him. The three-year-old toddler stood alongside a roadway in the dark hours before dawn on July 3, 1925 as an errant driver struck and killed his father who was waiting a tow for his disabled car. Floyd Jr. was also knocked to the ground. Floyd Sr. was only 25 years old, working for the Independent Truck Company and lived in Mount Vernon.

Floyd Jr.'s early years were apparently spent with his paternal grandparents Jacob and Carrie Emma Perin of Mount Vernon. Jacob was a real estate broker who had seen a bit of trouble for questionable speculation in forest land near Clallam, Washington in 1905. Evidence indicates that Floyd Jr.'s mother Nellie Juanita Humphrey remarried to Samuel J. Walker in 1926 in Seattle after her husband's death. Nell Walker later married Robin Carswell of Seattle in 1941.

Floyd initially attended grade schools in Mount Vernon until his grandparents moved to East Stanwood where he went to Lincoln High School. His grandfather Jacob purchased a small farm in the area. By 1940 Floyd was working in a retail bakery shop. Perhaps unsatisfied with the direction of his life, he dropped out of high school and enlisted in the Army on January 15, 1941. Floyd would be assigned to Company B of the 65th Engineer Combat Battalion in support of the 25th Infantry Division known as the "Tropic Lightning" Division. The motto for the 65th was "First In, Last Out." Part of the 65th deployed from Fort Lewis, Washington. Dormant after World War I, the unit was reactivated January

1, 1944 and was engaged in improving defenses around Oahu, Hawaii when Pearl Harbor was bombed. The 65th was a multi-faceted unit which fought when called upon and was responsible for clearing the way for other American troops. They built bridges and trails, repaired roads, built or destroyed obstacles to ensure a supply route for advancing troops. They also were responsible for clearing any booby traps and land mines planted by the Japanese.

They were quickly deployed to the South Pacific where they saw action on Guadalcanal in the Soloman Islands and saw intense combat in the rainforests of New Georgia. When the 25th Division landed on the Philippine Island of Luzon on January 11, 1945 the 65th Engineer Combat Battalion including Pfc. Floyd Perin was with them. They would be in continuous combat until taken off the line on June 30, 1945. The 25th Division which included the 27th and 161st Infantry regiments pushed toward the vital railhead at San Jose as the American Sixth Army battled toward Manilla. Blocking their way was Japanese General Tomoyuki Yamashita the "Tiger of Malaya," who had taken Singapore in February 1942.

General Isao Shigemi was responsible for defending the stronghold of San Manuel with over a thousand men and roughly 45 dug-in tanks. American troops began securing surrounding high ground on a ridge above San Manuel and decimated a counter-attack by the Japanese on January 16 and 17. The final push against San Manuel would come on January 24th and the town would fall four days later. It was the first significant tank engagement of the campaign. It is considered a minor engagement in the larger picture of liberating Luzon but it was the final fight for Floyd Perin who had seen much in his four years in the service.

Reports say that on January 21, the day of Perin's death, a probe of the Japanese lines took place. The enemy had been prolific in planting land mines throughout the area. One of them is blamed for causing Perin's death. Medical records indicate he was killed in a vehicle after striking a mine; the force viciously ripping into his body. The three companies of the 65th Engineer Combat

Battalion all received Presidential Citations for Gallantry for their efforts on Luzon. Luzon was the last campaign for the Battalion in the Pacific during World War II. The baker's apprentice from East Stanwood was buried in 1949 in Evergreen Cemetery in Seattle. He was 23-years-old.

# Home Front Tragedy
## Orville Knutson

*Orville Knutson*

Orville Knutson worked for the Lien Bros. Packing cannery when he registered for the draft on October 16, 1940. He was 27 years old. Orville enlisted early in the war on April 4, 1942 less than four months after Pearl Harbor. He spent two years in the Pacific Theater of the war before being re-assigned to Camp Swift just east of Austin, Texas on the flat lowlands of Bastrop County. The Camp was built in 1941 as a military training facility which included small arms, weapons firing, combat engineering and other infantry skills. It is likely that Sergeant Knutson was an instructor at the camp as his obituary touted his efficiency as an American soldier: Asiatic-Pacific theatre ribbon, good conduct medal, sharpshooter's medal and expert medal with a light machine gun along with a bronze star. Efforts by this writer to discover more about his first two years in the Army have been unsuccessful.

Orville Peter Knutson was born November 16, 1913 in Warm Beach just south of Stanwood, Washington but soon moved with his family to Livingston Bay on Camano Island where the family patriarch Peter worked as a piler in one of the local lumber mills. In 1930 Peter found work for himself and his son Clarence with the local Stanwood electric company as a laborer and lineman. The five-foot eight, blond, blue-eyed Orville graduated from Stanwood High School in 1933.

News reports do not tell exactly what happened to Orville, just simply that there were complications following an operation. His death certificate issued by Bastrop County, however, provides some telling information. Sergeant Knutson was dead on arrival at the base hospital after an accident at 3:45 p.m. on February 7, 1945. Cause of death was given as "penetrating shrapnel wound lacerating aorta and pulmonary artery producing massive hemothorax and hemopericardium." The explanation suggests a tragic accident perhaps during a training session for a soldier who thought himself safely home.

# War Darkens a Local Home
## Vivien Mickel

*Vivien Mickel*

World War II brought the Mickel family to Stanwood. Edward Albert Mickel did not have deep roots in Stanwood, but the talented carpenter, machinist and businessman saw an opportunity. On March 17, 1945 he opened the Mickel Welding & Machine Works along Stanwood's waterfront. The building, once home to earlier efforts at oyster production, had his business on the first floor with his family comfortably located in an apartment upstairs. That family included his wife Leona, daughter Annabelle and son Richard who entered Twin City High School.

Another son, Vivien Donald Mickel was in training at Fort Lewis after being drafted in May of 1944. A welder and flame cutter himself, Vivien visited Stanwood and his family in December of 1944. As his family settled into their new community, Vivien was being shipped to the South Pacific where, as part of the 12th Cavalry Regiment, he would take part in the liberation of the Philippine Islands.

Al Mickel's business was touted as "one of the best appointed machine and welding shops in the northwest Washington, equipped to do anything from the fixing of a wheelbarrow to the finest precision work..." He was neatly situated next to the Stanwood Shipyards which had a lucrative contract to supply the U.S. Army Transport Service with grand scale barges. With the first launch on July 17, 1943, manager and owner Horace Kelsey would oversee the production of seven barges until mid-1945. In December 1944, coinciding with the Mickels' arrival, Stanwood's shipyards converted from wooden barges to a steel yard operation.

The invasion of the Filipino island of Luzon occurred January 9, 1945. It is likely that Vivien Mickel joined his regiment as a replacement after they had captured Manila and liberated the prison of Santo Tomas. Assigned as a scout, Donald would have been involved in cleaning out pockets of resisting Japanese troops within the Filipino capital. On April 2, 1945 Private Mickel was killed by an enemy sniper. The *Twin City News* proclaimed that war had darkened the Mickel home.

Soon after his son's death, Al Mickel got word that the Army contracts for Stanwood constructed barges was ending. The town's shipyards had served their purpose. Throughout the rest of 1945 and the following year, Al Mickel ran ads selling lathes, drill presses and boring machines as he terminated his Stanwood business. By the early 1950s, he was again working for others as a machinist in Edmonds.

Stanwood was a brief episode in the lives of the Mickel family but marred by a sense of loss—a promising business venture ended by unfortunate timing and a place forever associated with the telegram announcing the death of their son.

# "Pushing Forward"
## Daniel Hess

Danny Hess was remembered by some as the paperboy hocking news on the corner of Broadway and Hewitt in Everett or the worker for the Alexander Printing Company. Others knew the teen who played Principal Smudgely in the Stanwood High School play of 1943—the same year he earned his letter in football. It was also the year that Danny, in his junior year, volunteered for the U.S. Marine Corps. The last time Danny Hess volunteered for anything was on May 24, 1945 during the battle for Okinawa, Japan.

Daniel's father Joseph Hess was a typical journeyman laborer moving from farm to factory in Washington. His son Daniel Sylvester Hess was born in Yakima on June 19, 1925 after

*Danny Hess*

Joseph found work in the area's agricultural fields. Joseph found steady employment in the Everett lumber mills by 1930. Danny's maternal grandparents, Carl and Caroline Satra, ran a dairy farm near Florence. Following his grandfather's death in August 1940, Danny lived with his grandmother while he attended Stanwood High. Like most youths, say relatives, he sought to escape from the quiet life of a small community and the routine toil of farm work. He longed for excitement. What could be more exciting than traveling overseas in the great crusade to destroy world fascism. After joining the Corps, he trained at Camp Pendleton and San Diego before being assigned to the First Separate Engineer Battalion. Their mission was to fight along with combat troops and maintain vital routes of communication, construct airfields, bridges and camp facilities under the most adverse and dangerous conditions and perform demolitions, mine detection and bomb disposal.

During his 19 months overseas, Danny saw much of the worst the Pacific Theater of War could provide, including the carnage of Iwo Jima. Okinawa, however, would prove the bloodiest battle of the Pacific campaign. The torrential rains began in mid-May 1945 making it treacherous for heavy equipment such as tanks to move forward across the island. The intense and fortified Japanese resistance took its toll; soldiers often remaining where they fell in a grotesque embrace of death. "Hell's own cesspool," remembered Private Eugene Sledge. Losses included dozens of supporting Sherman tanks, many disabled by the thousands of landmines planted by the Japanese. Engineers had the duty to clear those mines for the advancing Shermans. In one case, when volunteers were sought for a particularly hazardous job of going ahead of a tank column to defuse explosives and unexploded shells, Danny Hess stepped forward. Severely wounded that day, he died two days later.

Danny's battalion received a unit citation for their efforts in the Pacific. It read in part: "Undeterred by both mechanical and natural limitations, the First Separate Engineer Battalion completed with dispatch and effectiveness assigned and unanticipated duties which contributed immeasurably to the ultimate defeat of Japan and upheld the highest traditions of the United States Naval Service." One who upheld those traditions was a former Everett newsboy and student from Stanwood, Washington. Danny Hess' obituary stated that "he died as he lived—pushing forward."

# "Victims of Perdition"
## Edwin Lund

*Edwin Lund*

It was a clear night as the Boeing Superfortress K-43 lifted into the calm evening sky from Guam on April 13, 1945. Other planes from the 330th Bomber Group followed at one-minute intervals. Nearly fifteen hundred miles later it would join in the largest fire-bombing attack on Tokyo to that date in the war. There were eleven men in the crew including 21-year-old Edwin Palmer Lund whose hometown was given as East Stanwood. It was only the second mission for this crew which was accompanied by a ranking observer, Lt. Colonel Doyne Turner, commanding officer of the 458th Bomber Squadron, of which the 330th was a part. Some reports cite the plane's nickname as the *Weddin' Belle*.

The 16 planes from the 330th BG would fly individually until meeting up with other squadrons off Japan where a lead plane would guide them to the target area. There would be 327 B-29s involved in the raid that night, all armed with high explosive and incendiary bombs. It was classified as a precision night mission whose code name was "Perdition 1." Their target was the Tokyo Arsenal Complex, the greatest such complex in the Japanese empire. The attack would last roughly three and a half hours with the last bomb dropped at 2:26 a.m. on April 14th. Edwin Lund was the tail gunner on the K-43 (tail marking designation) and, as a private first class, was the lowest in rank.

Edwin was born in Seattle on November 12, 1922, almost exactly nine months after the marriage of his parents Esvald and Emma Haugen in Snohomish. A year later the family lived in Edmonds where their second son Theodore was born. Esvald immigrated to America at the age of 18 in 1898 settling first in Sault Ste. Marie, Michigan. As with many immigrants, Esvald took what jobs he could find. By 1918 he was working in a shingle mill in Everett and would stay in the mills for years except for a stint as a fisherman in 1935. Both sons later expressed their love of fishing. Finding a piece of land of his own, however, must have been important to Esvald since by 1940 the family had again relocated to a small dairy operation a few miles east of East Stanwood in an area designated as the "Village" precinct on voting records, near Cedarhome. The boys attended Stanwood schools for a short time. Esvald kept the property into at least the 1950s.

The B-29 Superfortress was a marvel of American technology developed by Boeing starting in 1942, much if it at the company's plant in Renton. It was the second pressurized aircraft cabin mass produced in the U. S. (after the Boeing 307 Stratoliner) and that expertise would spur Boeing's great success in the post-war years. Unlike its Boeing cousin, the B-17, it also was heated and much less cramped for the crew with five electrically powered gun turrets. Even with the greater space on the Superfortress it would have been a tight squeeze for the five-foot-eleven, 153-pound Lund. It could cruise at over 31,000 feet with a top speed of 350 mph. This enabled the aircraft to fly above most of Japan's defense network.

Ironically, Edwin Lund had found work on the assembly lines at Boeing Aircraft Plant #1 in 1942, the same year he registered for the draft. He was joined on the assembly line that same year by his brother Theodore. We do not know if they worked specifically on the B-29. The great demand for labor to work the defense plants may have assisted keeping the boys on the homefront. According to scholar Polly Reed Myers, however, by mid-1942 the slowdown in production forced Boeing to change its policy of an all-male work force leading to the necessity of hiring women and people of color. Single men had a greater chance of being drafted such as Edwin Lund, who joined the Air Corps on March 24, 1944.

The B-29 started combat operations in early June 1944 with bombings beginning in earnest in November of that year. While initial high-altitude bombing using radar was yielding results against military targets, the new commander of the XXI Bombing Command, General Curtiss LeMay, believed low level firebombing would yield even greater results, increasing accuracy, and allowing for larger bomb loads and less fuel and weather problems. This, of course, exposed American air crews to the greater lethality of Japanese anti-aircraft flack and fighter planes. LeMay, however, was excited about the results of a ten-day firebombing campaign in March 1945. Richard Sams wrote that "this ten-day firebombing campaign, particularly the Great Tokyo Air Raid, transformed the status of the USAAF overnight. Intoxicated by these spectacular successes, LeMay and other leading air force officers were now convinced that firebombing alone could end the war." Low level firebombing would begin in earnest. Only one problem intervened: the Air Corps ran out of incendiary bombs, a logistical problem solved in mid-April just in time for K-43 and the massive raid on April 13.

There are no eyewitnesses to the last moments of the *Weddin' Belle*. Their specific target was the Tokyo Artificial Chemical Fertilizer Plant. We do know from other participants in the attack that the anti-aircraft fire was intense and the Japanese used rocket-propelled planes called Bakas which carried a bomb in their nose and were flown by Kamikaze pilots. Some were reported to have attacked the planes of the 330th. The squadron's only loss would be the *Weddin' Belle*. Records show that the crew was able to bail out of the dying plane. The aircraft crashed at Sanuma, Omiya Village about 40 miles northeast of Tokyo. Eleven of the twelve crew members were captured including Private Lund. The body of the plane's navigator, George Kruse Jr., was later found in a wheat field. The B-29s dropped over 2,000 tons of bombs on the raid, killing over 2,400 people and leveling over 170,000 homes. The attack did destroy Building No. 49 in the complex where Japan scientists were working on the development of an atomic bomb. However, the *Weddin' Belle's* efforts against the Artificial

Chemical Fertilizer Plant reportedly destroyed only five percent of the building.

The eleven survivors were turned over to the Kempei Tai—Japan's version of the Gestapo—and interrogated. They were then taken to the Tokyo Army Prison in Shibuya in western Tokyo where they were confined along with 51 other American airmen. They would become, wrote Sams, the other "victims of Perdition." The Great Yamanote Air Raid against Tokyo took place May 25-26 by 464 bombers, with participants again from the 330th Bomber Group. The planes dropped over 3,200 tons of bombs in what would be the fifth and last night of low level incendiary raids against Japanese urban areas. It would also be the last evening for the 62 airmen in the Tokyo Military Prison. The prison was accidentally hit by American ordnance during the raid and caught fire. The Japanese guards refused to unlock the cell doors leading to the death that night of all the American POWs. A tribunal after the war found the guards guilty of war crimes and ordered them executed. There are no gravesites to visit for Edwin Palmer Lund and his comrades, only their names which rest on the gray stone walls of the Courts of the Missing in Honolulu's National Memorial Cemetery of the Pacific.

"We left as boys and came back as men. Lets hope no one has to do it again. We took a plane brand spanking new, started as strangers and came back a crew. We were young and slim, our backs were straight. Our eyes now dim, we know our fate. We had to go, a job to do. Our friends all went, we had to too. Now years have passed and soon we'll rest. The whole world knows we did our best. Above the clouds our spirits will soar. When life is over, it's through the next door. We'll join up again on another plane, take off for the heavens, a crew again."

Anonymous, posted on Findagrave

# "A Damn Long Day"
## Arne Aalbu

Arne Aalbu was a full grown man of 31 when his family moved
from South Dakota to East Stanwood sometime around 1942—
just in time to be drafted into the U.S. Army for World War II.
His mother Serene Sneve Aalbu had family in the area. The
unmarried son of Norwegian immigrant farmer Iver Aalbu,
Arne worked as a fisherman and oysterman. He was assigned
to First Platoon, Company L, 134th Infantry--part of the 35th
Infantry Division. Their first assignment was as coastal defense
along California's shoreline. They received advanced training in
Alabama and North Carolina. After landing in England in late
May, 1944, they became part of the massive build-up of troops in
advance of the D-Day invasion of Normandy. Held in reserve, the
35th Division joined the fight in France on July 11, 1944.

*134th Infantry, Arne Aalbu, Second row from the top, Seventh from the left*

The beaches of Normandy had been tamed but the ancient, entangled maze of French hedgerows, all of irregular height and size, made slow and dangerous going for the Americans who complained of feeling entrapped in such a labyrinth of growth, never knowing what lethal force might meet you at any turn. Arne, now a staff sergeant, was wounded during this time after fighting to liberate St. Lo, a critical crossroads for the Allies.

During Arne's hospitalization, the 35th Division, part of George Patton's Third Army, raced across France. Arne rejoined his unit in November in northeast France as they prepared to cross the Saar River into Germany, the last major obstacle before the 390-mile string of enemy fortifications known as the Seigfried Line. This effort along the southern German border was in coordination with allied troops crossing the Rhine River farther north under command of English Field Marshal Bernard Montgomery.

The Saar crossing had its problems especially for tanks. There was no easy access, only a pair of partially intact railroad bridges permitting passage by troops but not vehicles. The gamble was to get all three battalions of the 134th over by foot before any counter-attack. The effort began at 5 a.m. on December 8, 1944. German machine gunners however, hidden in buildings near the river pinned down the last troops including platoons of K and L Companies. Those men did not successfully cross the river until late afternoon. During this time Arne was again wounded, this time seriously. He died from his wounds the next day—being one of the first Americans to set foot on German soil. Captain Raymond J. Anderson reported the loss of Arne Aalbu and two other soldiers during that day's fighting in his after action report. In reflecting on the day, Captain Anderson wearily noted that it was a "Damn long day."

# Death by Divine Wind

## Robert Harrison

*Robert Harrison*

As a Water Tender second class, Robert Frederick Harrison worked in the hot, cramped fire rooms located in the bottom and aft part of his ship, the U.S.S. *Reid*. He helped maintain the boilers and fires which produced the steam that powered the *Reid*, a Mahan-class destroyer, which had survived the attack on Pearl Harbor and was credited with shooting down one of the Japanese aggressors that day.

Robert's father John and his family immigrated from England in 1892. John married Thora Anderson in 1919 and worked as a farmhand for Carl Peterson near Stanwood. He and Thora had their own place in Cedarhome when Robert was born in June 1921. Robert enlisted in the Navy in July of 1942 after graduating from Arlington High School. In February 1943 he was assigned to the *Reid*.

The *Reid* had its fair share of patrol and support duties from the Soloman Islands to the landings in New Guinea. In November 1944 they steamed to Leyte Gulf to support landing craft moving men and material to beaches on Leyte in the Philippines following American landings in October. By early December Japanese air cover had been severely degraded, pushing the Japanese to rely on Kamikazes, a corps of suicide bombers meaning "Divine Wind" in Japanese. By December 11th, crew members responded to ten alarms a day for battle stations and tried to steal an hour or two of sleep when possible. Around five in the afternoon twelve torpedo bombers began their attack on the *Reid*. Two were downed by U.S. fighter cover, anti-aircraft fire took down four more. Five planes, although badly damaged, zeroed in on the *Reid*. One plane crashed at the ship's waterline, its bomb doing severe damage to the ship's forward. But the next one provided a fatal blow. It slammed into one of the *Reid's* gun mounts with the resulting explosion setting off the magazine in the ship's aft. Within two minutes the destroyer had rolled violently onto its starboard side, explosions opening its stern, and plunged to the bottom 600 fathoms below, taking 103 of its crew with it. Very little time for those men below decks, such as Robert Harrison, to escape even if he was still alive.

Wayne Haviland was one of the survivors. "We had a clear view as she [the *Reid*] slipped, bow up, under the Comotes Sea. That picture remains firmly in my memory. I never think of it without experiencing an overwhelming sense of loss and sadness, because we later learned of the virtual annihilation of the aft damage control party and both fireroom crews . . ."

The bodies of the men, including 23-year-old Robert Harrison, were lost in the dark waters of the Leyte Gulf. The *Reid* was the last destroyer sunk in that bay. Two weeks after the sinking of the *Reid*, all organized Japanese resistance on Leyte ceased—on Christmas Day, 1944.

# A Deadly Christmas

## Robert Bransmo

*Headstone of Robert Bransmo*

December 23, 1944 was a cold, wet evening on the docks of Southhampton, England. Over 2200 men of the 66th Infantry Division were being ferried to France to assist in relieving surrounded American troops at Bastogne—known as the Battle of the Bulge. Confusion began in the boarding process with units separated and housed in different parts of the *Leopoldville*, a Belgian passenger liner converted to a troop carrier. Confusion later sent roughly 800 soldiers to their death beneath the dark waves of the English Channel after two torpedoes from a German U-boat slammed into the rear of the ship in the late afternoon of Christmas Eve. An estimated 350 men died instantly.

In the fetid, close quarters below decks of the *Leopoldville* as part of C Company, 264th Regiment, was a 24-year-old Staff Sergeant from Stanwood, Robert Sigurd Bransmo. Four years earlier, Robert had graduated from Stanwood High School and was clerking in S. A. Thompson's clothing store when he enlisted in August 1942, five months after marrying Helen Carlson of Mt. Vernon. The bride wore a gown of pink taffeta—her only jewelry a zircon necklace from Robert. He was a naturalized American, having been brought from Canada in 1922 by his Norwegian parents Peder and Mary.

As the *Leopoldville* slowly sank by the bow in the darkening hours of Christmas Eve, its Belgian captain failed to radio for help from Cherbourg, just five miles from the attack. Crew members reportedly escaped the ship in what life boats were available. When the order finally came to abandon ship it was given in Flemish, leaving Americans on the *Leopoldville* unaware and many still believing a tow would take the ship into port.

The scene was chaotic by the time one destroyer escort, the HMS *Brilliant*, came alongside. Soldier Donald Shaub survived the attack:

"I grabbed a rope and waited until the *Brilliant* was coming up and swung out on the rope in the darkness. I let go out of the rope, dropped about six feet and fell on the deck of the *Brilliant*. When I was waiting to swing out my buddies told me, 'We'll see you on shore.' Finally the *Brilliant* had to pull away. It had taken on 700 American troops and *Brilliant* was being beaten badly by hitting against the *Leopoldville*. The damage was evident and the lines were cut and they pulled away. Within 30 minutes after they pulled away the *Leopoldville* made a big surge up and went down to the bottom."

Some soldiers jumped to the decks of the *Brilliant*, some fell into the waters between, overladen in their heavy wool greatcoats and helmets, and were either crushed between the ships or drowned. Others later died of hypothermia like Robert Bransmo whose death date is given as Christmas day. Several hundred men were saved by the *Brilliant's* heroic efforts but upwards of 500 soldiers perished with the ship. One survivor later remarked that none of his training did him any good that evening. Military authorities reportedly worked to withhold information about the tragedy with many records remaining sealed until the 1990s.

Robert was first interred in the cemetery at Normandy but finally brought home and buried in the National Cemetery in San Bruno,

California in 1949.  The site of the *Leopoldville* disaster is today a
designated war grave.

# The Last Shot
## Wesley Sigerstad

When allied forces crossed the Rhine River into Germany
in March 1945 the end of Hitler's Thousand-Year Reich was
numbered in days.  Soldiers, such as those with the 387th Infantry
Regiment, certainly hoped to avoid becoming a final, fatal statistic
in the waning hours of the conflict but areas of resistance still had
to be subdued.  The 97th Division, including the 387th Infantry,
was tasked with eliminating the so-called "Ruhr Pocket"— the
industrial area in southwest Germany centered around Dusseldorf
which provided much of the economic might that powered
the German war machine.  The 387th Infantry was ordered to
secure two key bridges across the Wupper River, southeast of
Dusseldorf.  Pfc. Wesley Sigerstad of East Stanwood would be part
of that effort.  His regiment was involved in a pincer movement
to enclose and slowly squeeze the "Pocket" until its surrender.
The battle for the "Ruhr Pocket" would mark the last organized
resistance by German troops on the western front.

A year earlier in 1944, Wesley, known as "Sig" to his friends,
was attending East Stanwood (later Lincoln) High School.  The
17-year-old volunteered for the army on March 22, 1944, a month
following a student strike at the school against Superintendent
Alfred Tunim.  Whether that influenced young "Sig" to leave
school early is unknown but the timing is curious.  "Sig" shipped
out for Europe in February 1945 after training in California.

In the murky darkness just after midnight on April 16th, the
387th, part of 1st Battalion, moved toward their objectives
through dense, wooded terrain which gave cover to stiff German

resistance at times. However, they secured the river bridges intact at Wuppertal and Solingen by noon, advancing the encirclement of over 300,000 German troops in the "Pocket." Many in the 387th, and in particular Company B, received Purple Hearts for their efforts to secure those bridges. So did Wesley Sigerstad, posthumously. The accompanying letter to his parents from the Secretary of War noted the "slight intrinsic value" of the medal, but rich with the tradition for which so many gave their lives.

Wesley's parents received a second letter from the chaplain assigned to his regiment. Wesley "travelled with us for many a mile and his spirit will go on with us to bring this war to a successful conclusion," the reverend wrote. "He performed his duty courageously and well."

Dusseldorf and the "Ruhr Pocket" fell on April 18, two days after Wesley was killed. Three weeks later on May 7, 1945 — the day of Germany's surrender — Pfc. Domenic Mozzetta of Wesley's company was credited with firing the last official round in World War II's European Theater of Operation. Private Mozzetta's action is memorialized on a monument at Ft. Benning, Georgia, dedicated in 2000 — a tribute to all, including Wesley Sigerstad, whose last shot went unrecorded.

# The Peace Medal
## Richard Hiday

Lieutenant Richard Hiday was awarded the Bronze Star for "heroic achievement against the enemy" in the field during April 1945. Not for lives which were taken, but for lives which were saved. Hiday was part of the spear-point of the U.S. advance through Germany and Austria in the last days of a crumbling Third Reich. His reconnaissance patrol from the 5th Regiment, on three occasions, convinced garrisons of German soldiers in

*Richard Hiday*

places like Kohlgrube and Ober-Ottnung, Austria to surrender their arms without a fight, avoiding any further and meaningless bloodshed. An Oregon newspaper reported that not only did he capture 380 prisoners, "but also cleared the desired route so that reinforcements for the battalion might be brought forward." The 5th Infantry were some of the first to make contact with Soviet troops at Steyr, Austria on May 8, 1945.

Much of Richard Nowell Hiday's early life was spent growing up in and around Salem, Oregon. His birthplace, however, was Stanwood, Washington on Sept. 19, 1914. By 1920 he was living with his grandparents in nearby Conway. Parents Harry and Nellie Hiday moved their family to Salem in 1929 when Richard was 15. He graduated from Salem High School in 1933 and worked for a paper and pulp company until enlisting in the army in July 1942. Richard's fast advancement through the ranks points to a person with natural leadership instincts. Initially Corporal Hiday was assigned to a coastal artillery unit but was made a second lieutenant while in training in North Carolina. Re-assigned to the infantry, he was promoted to first lieutenant at Fort Benning, Georgia and shipped to Europe in late January 1945 as part of George Patton's Third Army.

Richard's death came seven months after the surrender of Hitler's Germany. As a member of the army of occupation, he was stationed near Munich, Germany with the 71st Division. On Christmas day, 1945, Hiday was detailed to bring wounded American soldiers from the camp hospital to the regimental barracks for a Christmas dinner. A collision with another Army vehicle, however, put an end to any celebration. Fog obscured the roadway that day and the accident left the 31-year-old officer dead. The drivers of the two vehicles were not charged.

Four years later, almost to the day of the accident, Richard Hiday finally came home—transported by the government at the request of his family. He was laid to rest a month later at the Belcrest Memorial Park in Salem. A newspaper stated that Richard Hiday's "persuasive manner" and diplomacy by engaging the commanders of enemy troops "caused them to surrender without a fight." His "heroic achievement" saved the lives of friend and foe alike. Richard Hiday had volunteered for combat as a warrior of virtue but came home a soldier of peace.

# Korea
## Containment of Communism

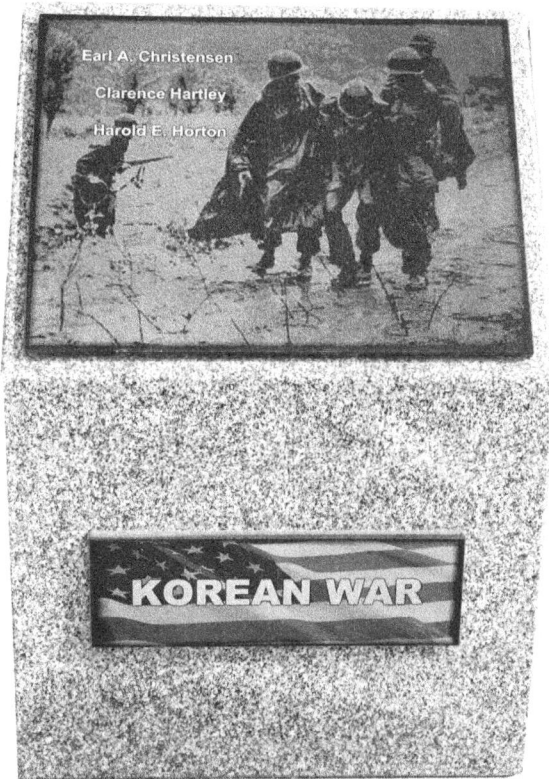

In 1945, the global threat of fascist oppression had been defeated and the world turned to rebuilding the shattered nations of western Europe and Asia. Peace for Americans would be short lived, however, as former allies turned to competitors for dominance on the new world stage. Korea had been divided at the end of the war with the north controlled by the Soviet Union and the south by the United States. The country was held in trust by the two powers with the public understanding that a unified Korea would be the eventual goal. Negotiations immediately became contentious and led to the creation of two nation states, the Republic of Korea in the south supported by the U.S. and the Democratic People's Republic of Korea backed by the Soviet Union.

The invasion of the south by North Korea on June 25, 1950 marked an effort to end the stalemate and make Korea one nation under the northern mantle. President Harry Truman reacted swiftly to the incursion and on June 27 announced that American forces would intervene to prevent a takeover of South Korea authorizing General Douglas MacArthur to command the effort. Harold Horton and Clarence Hartley were with the 1st Cavalry Division which landed in the Pusan region on July 18 followed five days later by Earl Christensen and the 2nd Infantry Division. Their labors and those of other American units stopped the near collapse of defensive efforts and held back the attacking North Koreans giving time for additional troops under the umbrella of the United Nations to begin driving the aggressors back north.

On December 16, 1950 President Truman declared a state of emergency to oppose "Communist imperialism" and called on "all citizens to make a united effort for the security and well-being of our beloved country." He asked "our farmers, our workers in industry and our businessmen to make a mighty production effort to meet the defense requirements of the nation." The citizens of Stanwood and Camano again heard the siren and responded. Local writer Alice Essex stated that meetings were held, volunteers recruited and observation posts were again established. Classes were begun in first aid and a blood bank was organized. The American Legion declared that "we can't wait until the first bomb falls" and mobilized a Civil Defense program.

Another local member of the 1st Cavalry was Donald Asmus of East Stanwood whose August 1950 letter to his parents gives some insight into the urgency of gearing up for another fight . He wrote, "when we left Japan I was an ammo carrier, but am now 2nd gunner and am learning 1st gunner's job. Because we are short in 1st gunners. Sure was hard at first but I have caught on to it now pretty well.  It doesn't take long to learn if someone is shooting back at you."


# Pure Survival
## Harold Horton

The day Harold Everett Horton died was a day of confusion and agonizing decisions by American commanders.  The North Korean invasion in June 1950 threatened to overrun the entire Korean peninsula.  American forces stationed in Japan, largely untested and with no battle experience, were quickly mobilized and inserted into the last toehold—a 5,000 square mile area in the southeast of the peninsula known as the Pusan Perimeter after the port it defended, along with the 1,400 miles of crucial rail lines which enabled the movement of troops and vital supplies.  One author called this defensive operation the "most complex mobile defense in U.S. military history."  Private Horton lost his life in that defense.

Born in Pasco, Washington on March 9, 1931, Harold attended schools in East Stanwood and Everett before enlisting in the 8th Engineer Combat Battalion of the 1st Cavalry Division.  Those who knew him remembered his "youthful exhuberance" which sometimes led to a bit of trouble for the young man.  Reportedly, the army was a judicial alternative for Harold after some joyriding in a car not his own.

His unit landed in South Korea around July 18th.  The engineers' job involved keeping open vital escape routes for retreating allied troops, demolishing those bridges which could be used by the enemy and manning a foxhole when needed.  The division rushed to establish a 35-mile defensive line just across the east bank of the Naktong River, a critical route for the northern invaders, especially the bridge at Waegwan.

*Harold Horton, Top row, Second from the right*

According to an investigation by the Associated Press years later, thousands of Korean civilians packed the roadways moving hurriedly south by foot and cart to escape the North Koreans. One soldier described them as "women clutching children, old men, overloaded ox carts." It was believed by American commanders that many in the fleeing crowds were North Korean soldiers dressed as refugees attempting to infiltrate their lines. There had been reports of disguised guerrillas opening fire on unsuspecting American soldiers. Confusion controlled the moment. Soldiers of the 14th Engineers took two days to wire 7,000 pounds of explosives to the steel-girder bridge. General Hobart Gay gave orders that the tide of frightened civilians had to be blocked from crowding onto the bridge. All action seemed futile including his men firing warning shots over the heads of the stream of people.

Finally as the light began to fade in early evening of August 3rd Gay later said that "there was nothing else to be done," and ordered "Blow the son of a bitch," one veteran remembered. Then Lieutenant Edward L. Daily recalled that "like a slow-motion movie," the massive explosion shattered the bridge supports dropping its steel spans and hundreds of Korean refugees into the muddy Naktong. Hundreds died when the bridge came down. Every man, woman and child on the bridge was killed, Daily remembered, and added that ten disguised North Korean soldiers were among the dead, although others did not recollect that fact. Another soldier who was there said "there was people on that bridge when it went up and during war that's the story. They're up there and they pull the plunger and that's it." Veteran Robert Russell said "I didn't like to do it, it was just pure survival at the time."

Panicked families now stranded on the far shore jumped into the river and attempted to swim across. Korean witness Kim Bok-jong told the AP that "many—I mean many—people drowned. Women and kids were exhausted before reaching the southern bank and disappeared under water. Sometimes kids were abandoned in the middle of the river." Three days later Gay sent boats across the river and brought 6,000 refugees over to American lines.

Harold Horton was nearby working to rescue a needed ammunition train that also crossed the Naktong at Waegwan when the order was given by General Gay to blow the structure. Harold died nearby. Reports indicate that the 19-year-old's death occurred after his truck was hit while leaving the area or perhaps in the chaos, he was mistakenly hit by the truck and killed since his injury was described as "crushing" due to traffic accident versus pedestrian. He was reportedly the Korean War's first known death from Snohomish County. AP writers Charles J. Hanley, Martha Mendoza and Sang-Hun Choe won a Pulitzer Prize for investigative reporting for this story.

# They Died Unknown
## Earl Christensen

In the fall of 1950 Nels Christensen walked into the offices of the *Stanwood News* to renew his subscription to the paper. The Anacortes resident had once lived at Camano Island's Tyee Beach and still kept abreast of the news of his former community. Nels might be called a journeyman laborer traveling to whichever area had an available job. Nels had served in the U.S. Army during the Great War only seven years after migrating to this country in 1910 from Denmark. He worked the mines and the railroads around Whitefish, Montana where he married Hazel Hooper in 1923. Their son Earl Anson Christensen was born in April a year later.

After reaching the age of 18 Earl registered for the draft of the second global conflict in 1942. At the time he was a farm laborer for Harry Pease of Marysville. Six-feet tall with gray eyes, the young man was assigned to the 254th Infantry Regiment of the 2nd Division and trained at Camp Van Dorn in Mississippi. Many of the regiment were trained and reassigned to other units. The main group of the 254th landed in Marseille France in December 1944 and became part of the 63rd Division of the 7th Army. It would see ferocious fighting in the last months of the European conflict smashing through the Siegfried Line in mid-March 1945 before being pulled from the front in April.

The unit was deactivated in September 1945 but Private Earl Christensen decided that he was not through with the army and re-enlisted October 24, 1945 and became an active reservist. The private was now made a Master Sergeant. His obituary states that he became part of the occupation forces in Japan soon after but when hostilities erupted on the Korean peninsula in June 1950, he was part of Company D, 9th Infantry Regiment, 2nd Division which was headquartered at Fort Lewis in Washington. Elements of this Division were the first to land at the port of Pusan in the southern most part of Korea toward the end of July and immediately were thrown into stopping the advancing North

*254th regt., Earl Christensen, Third row, Second from the right*

Korean forces. The 9th Infantry's first major engagement would be the bloody First battle of Naktong Bulge. It would be the last fight for Earl Christensen. The 9th had an assortment of older World War II vets who re-upped in the reserves along with many untested reservists.

Communist forces had been ordered to crush the Pusan Perimeter by August 15th, the anniversary of Japan's surrender, and casualties were not to be a consideration. The Pusan Perimeter was the last toe-hold for South Korea and its allies. On the night of August 6, 1950 a regiment of the North Korean People's Army (NKPA) crossed the Naktong River catching American forces by surprise. They seized high ground along Cloverleaf Hill and part of the Obong-ni Ridge and fortified the ridge bringing across more men, tanks and artillery over the next night and day. The 9th Infantry Manchus (a nickname acquired during the Boxer Rebellion) was deployed to the area on August 7th and elements were ordered to counterattack the afternoon of August 8th to regain control of Cloverleaf and Obong-Ni Ridge. Intense heat and the lack of available water took a toll on the Americans sapping their strength and adding to the difficulties of advancing.

That night the North Koreans re-took the ground seized by the 9th. Christensen's battalion had been placed on the left flank along the road to Yongsan. This see-saw action of attack and counter-attack between the 9th and other American units and the NKPA continued for several days until August 19th. Heavy casualties were incurred on both sides with the first two battalions of the 9th Infantry depleted by a third. By the end of this action, the North Korean 4th Division was annihilated for all practical purposes and what was left retreated back across the Naktong. In the end, the 9th infantry suffered 180 casualties including 57 killed—part of the roughly 600 Americans lost in this first battle of the Naktong Bulge. Twenty-six-year-old Master Sergeant Earl Christensen survived the harrowing and deadly action only to die of a heart attack on August 24, 1950.

Writer Jack Walker later recorded that early on "the fighting was fierce. U.N. forces were holding on by the skin of their teeth. New units arriving in Pusan were quickly thrown into battle…. The Army troops of the 24th and 2nd Infantry Divisions had fought their hearts out for eleven days in this area and stopped the enemy advance, but were too weak to push the North Koreans back across the river." The American action had preserved the Pusan Perimeter and turned the tide of the invading North Koreans. American reinforcements would now begin pushing the Communist forces back north starting with the amphibious landing at Inchon on September 15th. The battle of the Naktong Bulge was a turning point in the Korean War.

Lt. William R. Ellis, who experienced combat in World War II, says the 9th Infantry Regiment fought magnificently. "The original group of officers was gallant and far under-ranked. Most of the company commanders were (only) first lieutenants, which was a disgrace itself. They were forty-year-old, gray-haired World War II veterans (reserves called up) and still lieutenants in combat in 1950. I knew them all and have regretted at times that I did not join them (in death) for they by-and-large died unknown and unrewarded for their bravery."

According to Roy Appleman, "Few Americans today, or even then, know of the desperate struggle, the pain and suffering, the utmost heroic effort and valor displayed to stop the North Korean assaults. The U.S. suffered its highest casualties of the entire war during these six weeks. If and when the public does become conscious of this all-important battle, it will, no doubt, be ranked alongside Bunker Hill, the Alamo, Bataan, and Corregidor."

Master Sergeant Earl Christensen, who had survived some of the hardest combat subduing the German fatherland, saw combat in the Korean conflict for 17 days before his heart gave out. Although coronary thrombosis is written on official records as the reason for Earl's death, it does not do the persistent warrior justice. His body joined his father in Mt. Vernon's I.O.O.F. cemetery. He was only 26 years old.

# "Don't Worry About Me, Mom"
## Clarence Hartley

As a boy Clarence Hartley moved around a good deal. His father took jobs where he found them, usually in lumber mills or warehouses. Born in Clear Lake, Washington on January 6, 1933, Clarence attended grammar schools in Mt. Vernon, nearby Burlington, Everett, Stanwood and Napa, California. After graduating from East Stanwood's Lincoln grade school in May 1948, he spent his freshman year at Twin City High School before the family moved back to Clear Lake. Dissatisfied at his new school in Sedro-Woolley, Clarence dropped out and enlisted in the army on March 15, 1950 receiving his basic instruction at California's Fort Ord. Clarence called his mother in September 1950, who by this time was living on Camano Island, before being

*Clarence Hartley*

sent to Korea and the bloody conflict which took his life. He told her, "Don't you worry about me, Mom, just always remember that I'm coming home," he said.

South Korea's communist neighbor to the north invaded on June 25, 1950. A United Nations' force led by American troops deployed in support of the South Koreans. Pfc. Clarence Hartley was assigned to C Company, 7th Cavalry, 1st Battalion, 1st Cavalry Division. Trained as a bazooka gunner, he was sent to the front lines September 12th where his unit was assigned to support a tank battalion. The war and initial allied success changed dramatically on October 25, 1950 when thousands of Chinese troops crossed the Yalu River to rescue their faltering client state. South Korean forces buckled under the swift advance and numbers of Chinese forces jeopardized withdrawal of the U.S. 8th Army down the Korean peninsula. The job given the 7th Cavalry was as a rear guard to protect the Army's extraction before they were cut off.

Numbing temperatures settled on the regiment the night of November 29th as it dug in on frozen, snow covered, high ground alongside a critical road junction just north of Sinchang-ni. Initial probes by the enemy were beaten back by Company C but near midnight outposts mistook approaching Chinese troops for Americans. Suddenly the air erupted with the blare of bugles and whistles as Chinese attackers drove a wedge between two battalions, overrunning two command posts. However, the division and the 7th regiment regrouped and held. The fighting that night and the next day took 42 American lives with 140 wounded and five tanks destroyed. The Chinese reportedly lost over 1600 men. When Lydia Hartley returned home from the Camano Chapel on December 21st, a telegram waited, brutal in its simplicity—the government regrets. Just a month earlier she had sent Clarence a cake and presents to mark the 18th birthday he never saw.

Battalion Surgeon Capt. John Rourke remembered that some of the most frantic hours of his military career came during the attack at Sinchang-ni. We don't know, but it is entirely possible that the captain worked to save those cavalry troopers who poured into his aid station. One face he may have looked into, in the early hours of November 30th, belonged to Pfc. Clarence Hartley.

*Twenty-One Gun Salute at the dedication of the memorial in 2018*

# Vietnam
## Countering the Dominoes

Korea has been called the forgotten war. The conflict in Vietnam by contrast would long leave scars on the American body politic and its culture. On September 2, 1945, as the surrender of Japan was being accepted aboard the U.S.S. *Missouri* in Tokyo harbor, Ho Chi Minh was reading his Declaration of Independence before a mass gathering of half a million Vietnamese in Hanoi. The country's revolutionary leader, who had embraced communism while living in Paris (1917-1923), invoked the words of America's guiding document penned by Jefferson: "All men are created equal...." He had worked with American covert operatives during World War II in fighting the Japanese and sought American support in securing their rights of self-determination in post-war Vietnam. The Truman administration, however, was not going to upset a French ally which immediately asserted its rights of colonial ownership to that part of Indochina and the Vietnamese entreaties for help went unanswered.

France, which had controlled the northern provinces of Vietnam since its victory over China in 1884-1885, sent troops, with British assistance, to again occupy the region by the end of September 1945. Ho Chi Minh, and the Viet Minh, then began an active guerrilla campaign against the French, the culmination of which was the French defeat at the battle of Dien Bien Phu in the spring of 1954. Following the Geneva Conference the country was separated at the 17th parallel with the Democratic Republic of Vietnam and Vietnamese Communist Party controlling the north while the French ceded its power to the new state of Vietnam with Saigon as its capital in the south. Under these accords, a nationwide election was to be held in 1956.

The Eisenhower administration, fearing the spread of communist insurgencies around the globe after the world war, was determined to stop a communist consolidation of Vietnam and backed Ngo Dinh Diem in the south who, by late 1955, had established control over the government with American support. The U. S. began supplying military training and weapons. Both the Americans and northern communists remained active during this period in covert activities meant to strengthen their respective positions. Diem cancelled the nation-wide election declaring himself the president of the Republic of Vietnam. By 1957, the communists, now called the Viet Cong, began a program of terrorism in the south.

By the time the Kennedy administration began in 1961 the Domino Theory of nations falling victim to the communist threat was fully developed. Reports that the Diem regime was failing in suppressing the communists due to its often self-serving and corrupt practices, led to an American military report calling for an expansion of advisers, training, equipment and at least some number of American combat forces. While cautious, the Kennedy administration did sanction an increase from 800 American personnel in the 1950s to 9,000 by 1962. Political and religious tension, particularly with the Buddhist community in South Vietnam, grew over the summer of 1963. The U.S. administration maintained neutrality when South Vietnamese generals overthrew and executed Diem in early November 1963. John Kennedy was assassinated three weeks later and the problem of Vietnam became the property of Lyndon Johnson. Johnson's aggressive domestic programs, such as the war on poverty, would suffer with the escalation of American involvement in Vietnam. Johnson, like Kennedy, did not want the political backlash from watching yet another country fall to a communist plague.

Incidents on August 2 and August 4, 1964 in the Gulf of Tonkin would open the flood gates in the American fight to keep Vietnam in its sphere of influence. Reported clashes between North Vietnamese gunboats and U.S. destroyers rallied American public opinion and the following Gulf of Tonkin Resolution gave Johnson the discretion to conduct operations in the country as he saw fit. On March 8, 1965 the first marines waded ashore at Da Nang. By 1968 American combat forces would number 536,000. The transitional year of 1968, however, was a breaking point. The Tet offensive, beginning in late January of the year, while an American strategic victory, shocked Americans. The scope and breadth of capability demonstrated by Vietnamese communist forces clashed with the information coming from their own government. Media coverage and the public's attitude dramatically shifted. Americans increasingly turned against continuing the conflict. All of this coupled with the assassinations of Martin Luther King and Bobby Kennedy and finally the bloody Democratic Convention in Chicago. Johnson did not seek a second term of office in 1968.

In his 1971 memoir, Johnson wrote: "I knew from the start that I was bound to be crucified either way I moved. If I left the woman

I really loved – the Great Society – in order to get involved with that bitch of a war on the other side of the world, then I would lose everything at home."

# Outside the Wire
## Warren Gary Peterson

*Warren Gary Peterson*

First Battalion, 327th Infantry, 101st Division, Airborne was the first to deploy to Vietnam in 1965. It would not return home until April 1972. Their Headquarters Company would include the famed and the infamous.

Pfc.s John Sergeant Voegtli of Connecticut and Warren Gary Peterson of Stanwood were also part of that Company. Together, they lost their lives during combat overnight on February 9 or 10, 1966.

They were not part of that Company's notorious "Tiger Force," a special platoon created by Major David Hackworth trained to "out guerilla the guerillas." The elite unit carried out special reconnaissance and commando functions, blending into the jungle, seeking the enemy on its own ground. Hackworth would later be the inspiration for Colonel Kilgore in Francis Coppola's *Apocalypse Now*. The U. S. Army later investigated Tiger Force for alleged war crimes. Elements of the 327th, including Tiger Force, were involved in intense, bloody, hand-to-hand fighting on

February 7th in the Tuy Hoa Valley near the village of My Canh, as part of Operation Van Buren. At least 26 Americans were killed.

Warren Gary Peterson was 18 years old when he arrived in Vietnam, beginning his tour in November of 1965. Born in Mount Vernon in September 1947, Warren attended grade schools in Conway and Stanwood High School until transferring to Mount Vernon in his junior year. His brother William later graduated from Stanwood High. Their father Wilhelm had been a rancher in Montana when he married Emma Ruff, a beauty shop owner in Marysville. Warren enlisted in June 1965 right out of high school and trained at Fort Ord, California and Fort Benning, Georgia before being assigned to the 101st.

A defensive perimeter was established after the battle of My Canh. On February 10, 1966 U.S. newspapers reported that the night before, the battalion was fiercely attacked three times by Viet Cong; fighting so confused that soldiers complained that it was impossible to know who was firing at whom. Major Hackworth later wrote of those attacks on his camp's perimeter and of recovering two mortally wounded American soldiers from an outpost just outside the wire. "I called out to the two men on the outpost but there was no reply….We snaked our way out of the perimeter until we came upon the OP. It had been overrun, and a still-lit cigarette glowing in the darkness beside two very still bodies." Hackworth blamed the glow from the cigarette for exposing the outpost's position. Both were dying from multiple shrapnel wounds. Hackworth and another soldier grabbed the mortally wounded pair and hurriedly brought them inside the defensive wire to the aid station but it was too late. He does not name the soldiers but their shrapnel injuries were consistent with those listed on casualty reports for Voegtli and Peterson, the only two company soldiers listed as dying at that time. Vietnam researcher Bruce Swander believed that Peterson and Voegtli had less than 15 days left on their tour.

Warren Gary Peterson's body was buried in the Fir-Conway Lutheran Cemetery where his father joined him two years later.

Warren Peterson was the first known casualty from the Stanwood area to die in Vietnam. Steve Good had been an elementary school classmate of Warren's in Conway. He posted his thoughts on learning of his death. "I can still remember the sadness and shock I felt," said Good, "When I look at his senior picture today it saddens me deeply that Warren missed so much of life. God Bless Warren..."

# A Sky Soldier Falls
## Frank Smith

Friends of Frank Lee Smith remember a young man who never shied away from a fight and perhaps was a little too eager to begin one. On January 16, 1967, Frank certainly found himself in the fight of his life—a lethal one for the 25-year-old soldier as it turned out.

Although born in Elko, Nevada in 1941, Frank lived on Camano Island from the age of three months. He attended Twin City High School as had his father Carl. Carl worked for several years for Twin City Foods in Stanwood. Frank's family was part of the history of the local area. His grandfather Louis Smith had received farmland on the southern part of Leque Island from his uncle and foster father, Nils Paeter Leque, who came to the Stanwood area around 1877. Later the family owned property on Camano Island along what is today Smith Road.

Frank joined the U.S. Army in June 1965 and began his tour in Vietnam a year later. He served with Company B, 503rd Infantry of the 173rd Airborne Combat Brigade, 4th Battalion—dubbed "Sky Soldiers." The battalions of the 173rd led Operation Cedar Falls in January 1967—an attack against a stronghold of the Viet Cong. The so-called "Iron Triangle" was a 60-square-mile area twenty miles north of Saigon composed of fortified bunkers

*Frank Smith*

connected by a massive honeycomb of tunnels which offered escape and protection for the enemy. News reports stated that "some of the tunnels had as many as five levels," containing weapons, ammunition, food, medicine and clothing. It was the largest assault in the Vietnam conflict up to that point utilizing 30,000 troops, intensive air bombardment and widespread use of the defoliant Agent Orange.

Frank's company was the first to make contact with the VC and assaulted a complex of bunkers. Several casualties were taken during the intense firefight. Lt. Daniel Severson regrouped the lead platoon and directed fire toward bunkers hidden in the dense foliage. As the Americans began to pull back, however, their right flank was hit by automatic weapons' fire from a concealed location. A grenade knocked Severson unconscious and killed radio operator Arthur Wilkie. Regaining his senses, Severson singly attacked and eradicated the gun emplacement. Refusing treatment for his wounds, Severson directed the withdrawal of the platoon, ensuring that none of his men, dead or wounded, were left behind. For his actions Lt. Severson was awarded the Distinguished Service Cross. The Triangle claimed the lives of four Sky Soldiers that day including Private Wilkie and Private Frank Lee Smith.

Operation Cedar Falls was praised as a success by military authorities who cited the capture of a large cache of weapons and explosives and destruction of the tunnel system. U.S. deaths were placed at 72 during the campaign. Much of the enemy had melted away into the jungle but their losses were put at 750 killed. However, six months later the Viet Cong again dominated the Iron Triangle and in January 1968 used it as a staging area for an attack on Saigon during the Tet Offensive.

# "A Day of Extreme Sacrifice"
## George Broz

Operation "Badger Tooth" began December 26, 1967. Its purpose was to search out and destroy enemy units known to occupy seaside hamlets in Quang Tri Province, South Vietnam. That deadly stretch of northern coast, along the country's Route 1, was dubbed by French troops the "Street Without Joy" after bloody engagements there in 1953. Christmas 1967 would be the last one experienced for a young lieutenant from Stanwood, George Michael Broz, who died among the sand dunes of Quang Tri, a little over a month after reaching the war-torn nation.

*George Broz*

The 23-year-old officer was born in Tacoma in 1944 but attended Stanwood High School, graduating in 1962. After time in Community College and the University of Washington, he worked as an accountant for the Weyerhaeuser Company in Everett. Former teachers and classmates remember George as a good student and an aggressive athlete; loyal and kind hearted. "George was one of us," remembered classmate Jim Joyce. "He was aggressive in sports, pretty smart in the class, got beet red with embarrassment when teased, just a good guy."

But in October 1965, George responded to America's deepening involvement in the Vietnam conflict and enlisted in the United States Marine Corps. George's father had flown B-17 bombers during World War II and encouraged George to follow the honorable duty of military service. George's eighth grade teacher Lee Ayres said he thought George had a desire to please his dad, even though they clashed at times. Ayres admitted to being disappointed when George joined the Marines. "I think he wanted to prove he was going to serve the proudest military organization he thought we have and make his dad proud of him." Duty to country and family certainly served as motivation and coming just over a year after the misguided Gulf of Tonkin

*George Broz*

Resolution, Americans overwhelmingly favored sending more U.S. troops to that country's civil war.

George spent his first year at a base in El Toro, California. He returned to Washington to wed Camano Island resident Elizabeth Neale in December 1966 and then entered Officer Candidate School in Quantico, Virginia a month later. Graduating in April, George entered Vietnamese language school, graduating with honors on November 15, 1967, the same day he left for Vietnam as a 2nd lieutenant in Lima Company, 3rd Battalion, 1st Marine Division. In December 1967 the Battalion was stationed aboard the amphibious assault ship the U.S.S. *Valley Forge* on the northern coast of South Vietnam. He wrote Elizabeth on December 9th, their first year anniversary.

"I can understand your bitterness about us being apart—I don't like the loneliness and killing anymore than you do; however, every people (race, if you will) has the right to fight for their freedom and to enlist the help through treaties and agreements, of other countries if the enemy be such that they cannot stand

and face him alone.  Right now we are fighting Communism in Southeast Asia, nebulous term and concept that appears every day in the newspapers.  We are fighting so that the Vietnamese people will have the opportunity to choose their own government; so that government will have the chance to get on its feet and be able to serve the people as it should.  I have been through the rice paddies and jungles have seen dirty, naked and diseased half-starved children living in places that made the woodshed look like a palace; I believe that these children should have the right to an education and the right to earn their living without the threat of Communist aggression and/or domination.  To bring it down to a more personal level, I have seen Marines shot and maimed by booby traps and killed from both.  The day I took over my platoon, one of my fire team leaders was killed and one of my squad-leaders critically wounded from an M-26 grenade booby-trapped in a bush alongside a trail.  We are fighting so these young men (18-19-20-years old) shall not have died in vain.  Even to a more personal level, I would rather be here now and have it done with so that my son (God willing that we have one) will not have to come over here twenty years from now; or worse, so that he won't have to be fighting on the West Coast of the USA, because we gave up in Southeast Asia in 1967 or 1968.

You asked if I believe in what I'm doing, if I have a cause.  I let the preceding speak for its self.  I probably got carried away, but I do feel rather strongly about my convictions and want you to feel the same."

As part of Special Landing Force Bravo, Lima Company spearheaded the attack of "Badger Tooth," securing a landing zone on the beach for the remaining battalion units which were ferried in by helicopter.  North Vietnamese regulars were suspected to be in the area and Lima was tasked with clearing the villages of Thon Tham Khe and Trung An.  Resistance was light that first day and the Marines spent a quiet night.  The next day, however, Lima Company was again ordered to sweep the villages and the outcome was tragically different.

As Lima's lead platoon approached the edge of the village, they were met with withering and lethal fire from multiple machine guns, RPGs, mortars and AK-47s, taking heavy losses. They were "surprised by a well-sprung enemy ambush" wrote military historian Bradford Wineman. Force commander Col. Max McQuown "later reported that when the company was 25 meters from the hamlet, 'the lead elements of Lima Company were blown away,' resulting in much of the heavy casualties of the operation. A subsequent investigation reported the enemy had withheld its fire on 'all fronts until attacking units were drawn into the killing zones.'"

Wineman wrote that "first squad was shredded - caught crossing open flat sand, with no cover or concealment. Friends from second squad, who tried to reach them to help, were also killed. In just a few hours, the bodies of 136 young Marines littered the sand surrounding a small obscure village." Forty-eight of the battalion died that afternoon. Another 88 were wounded. Among the dead was 2nd Lieutenant George Broz. Besides a stash of weapons, 31 of the enemy were believed slain.

Unknown was that the NVA 116th battalion was now waiting, hidden within a network of underground tunnels which inundated Thon Tham Khe, supporting disguised ground level bunkers. Wineman wrote that, "After-action reports and official histories detail how the NVA had turned Thon Tham Khe into a virtual fortress, with elaborate defensive positions and bunkers all connected by a complex system of tunnels. Villagers reportedly confirmed that the enemy had prepared the village fortifications for nearly a year."

It is likely that George Broz died during this firefight since his wounds are described as "gunshot wound to the chest and multiple fragmentation wounds to the body from hostile rifle fire and mortar fire…" The company regrouped but a second assault took the life of Lima company commander Captain Thomas Hubbell and inflicted 25 more casualties on Lima and supporting Mike Company before they drove the enemy from the villages the next day. The wounded and dead were taken by helicopter back to the *Valley Forge*. Former sergeant Michael Thill remembered that "triage aboard ship was a nightmare of chaos and confusion. It was a grisly scene. Our dead were stacked at one end of the hangar deck. The wounded, on stretchers, covered the remaining deck space."

The 3rd Battalion relinquished control of the area six days later following a New Year's truce, returning to their ships offshore, earning the nickname of the "Suicide Battalion" after Thon Tham Khe. Wineman wrote that "the Marines of BLT 3/1 mourned the loss of their comrades and began to speculate about who would get the blame for such a botched operation.... Much of the fault was laid on McQuown, prompting rumors that he would be relieved. Many in the battalion gave him the moniker of "The Butcher" for the rest of the deployment.

An investigation of the operation was ordered by the brigade commander but it agreed that McQuown's evaluation that the "formidable construction" of enemy defenses accounted for the outcome of the battle and the significant losses. According to Wineman, "while many officers used lessons learned from the devastating battle, the Marines who fought in it recall the battle in a more solemn, introspective light. Sergeant [Robert] McDavid succinctly reflected in his diary: 'It was the most miserable operation. I've never been so miserable in my life.'"

Operation "Badger Tooth" became an asterisk in the annals of the Vietnam conflict and the men who died there merged with the statistics of war. For many of the battle's survivors nightmares intensified "excruciating memories" of the conflict and brought many tears, as Michael Thill described: "When reading the names of friends from the past, especially those from my squad, I frequently paused for composure. At the same time, I thought of all of those from 3/1 who fell on 'Badger Tooth'. By the end of the Memorial Service, when 'Taps' was played, there were many many tears throughout the room. It was evident that we still hold all of our fallen brothers close in our hearts. They will live forever in our memories and in our souls."

The name and memory of George Broz has been evoked recently through the George Broz Memorial Scholarship, presented to Stanwood graduates by American Legion Auxiliary Post 92. A fitting tribute for one hero who gave his all on the "Street Without Joy."

Do not stand at my grave and weep.

I am not there; I do not sleep.

I am a thousand winds that blow,

I am the diamond glints on snow,

I am the sun on ripened grain,

I am the gentle autumn rain.

When you awaken in the morning's hush

I am the swift uplifting rush

Of quiet birds in circled flight.

I am the soft stars that shine at night.

Do not stand at my grave and cry,

I am not there; I did not die.

Mary Frye – 1932

# Beyond the Call of Duty
## Elliott Peters

Elliott Peters' tour in Vietnam ended in January 1968. Instead of returning stateside, however, he re-upped for another six months—a lethal decision it turned out for the 21-year-old corporal. On January 30, 1968 the North Vietnamese launched a nation-wide offensive coinciding with the national holiday of

*Elliott Peters*

the country. The Tet offensive was the longest and most controversial battle of the war and eroded U. S. support for the conflict.

Elliott Lee Peters was born in Coos Bay Oregon, but gave his hometown as Moxie City, Washington. He lived for a time in the Stanwood area, probably due to his father's migrant agricultural work. His brother Jim remembered the carefree times of two mischievous teens growing up together. "He was the one that I most looked up to when I was a little boy and the one that I fought with when I was a teen. We played war and cowboys and Indians together. Chewed dad's Copehagen, ate oatmeal raw on the roof of the back shed, started fires with matches that we stole from the neighbors car and those were things we did before we were 10 years old." He remembered nostalgically "just being young guys together, growing up, sharing times that we could never share with anyone else."

Heroism seemed to come naturally to Elliott and Jim recalled the time his brother saved a little girl from drowning, rescuing her from the bottom of Lake Rimrock. "He was my hero and the hero of the day. I do not think he even thought, he just dove in and got her out. I never ever told him how proud I was of him for doing that." By re-upping for a second tour, Jim said his brother hoped to trim six months off his military service. There was a girl in California that Elliott hoped might start a new life with him. "I would like to believe that that he would have gone there and that she would have become his wife and they would have made a good life together....I miss him, every day," wrote Jim, "but I also know that he is with me every day and we share my life in everything that I do." He lamented the family's loss with the death of Elliott and the blow to his parents—"when one of us died we saw a little of them also die." He said that his father's heart and spirit were broken, even as his mother rallied to be strong for them all after the crippling news. "I … thank you for being the person that you were," said Jim.

In the Marine Corps, Elliott was part of a Marine field artillery regiment (B Battery); 1st Battalion, 12th Marines, 3rd Division. But on March 18, 1968, he was assigned as a Forward Observer with Echo Company, 2nd Battalion of the 3rd Marine Division directing fire at enemy positions as part of Operation Ford. Marine Paul Marquis wrote of the extreme value of the FO: "he follows behind his radioman, why? because if a bullet hits his radioman it has to go through his chest and the radio before it hits the FO . . ." The FO, along with the company commander and radioman, follow a few meters behind the point man ready at a moment's notice to convey exact grid positions to a distant artillery battery once the patrol comes under fire.

During the Tet Offensive a contingent of North Vietnamese regulars had moved south to support the assault on the ancient city of Hue and Operation Ford was a search and destroy mission to root them out. Also in Echo Company that day was Sgt. Dale Dye who won a bronze star for bravery during the action. As an actor after leaving the Marines, Dye is recognizable in films such as *Platoon, Saving Private Ryan* and *Born on the Fourth of July*.

Leaving Phu Bai airbase, the Marines neared the hamlet of Trung Phoung when small arms fire pinned them down. Dye later remembered NVA fortifications being hit by Marine 105 howitzers—the shells eliminating a deadly machine gun nest and destroying the bunkers. Elliott Peters had done his job coordinating the artillery that hit the entrenched North Vietnamese and took out a mortar site shelling American troops. During the intense firefight, however, Corporal Peters was hit several times by small arms fire and died instantly at the scene. After being returned to the U.S. he was interred at the Fir-Conway Cemetery.

Normally, according to one Vietnam veteran, a lieutenant or high ranking NCO would have the critical job of a Forward Observer and not someone of Peters' lower rank. It is a testament to the battle-tested Marine that his superiors trusted Corporal Elliott Peters with the lives of their men.

# "A Very Bad Day On The River"
## Francis Campbell

His name is etched into the black granite of Panel 34W, row 13 of the Vietnam Memorial. Francis Duncan Campbell died the day after his 35th birthday on January 16, 1969 — three days before he was scheduled to return home to the U.S. He had been in the U.S. Navy for 14 years. As an Engineman Second Class, he was assigned to the YFU-62, a yard freight boat, part of the "Brownwater navy" which plied the murky channel of the Cua Viet River in Quang Tri Province, Vietnam. The boats brought ammo, food and spare parts to the remote outposts which dotted hilltops along Route 9 from Dong Ha on the coast to Khe Sanh near the Laotian border. It was a gauntlet of fire for the men who served on these boats as North Vietnamese gunners made them constant targets.

Francis, known as "Sonny," was born in Everett, Washington in 1934 and received his schooling there. His father O.D. "Ossie" Campbell was a long-time mail carrier in the city but Ossie and his wife May later settled on Camano Island. Sonny's home of record at the time of his death was Gearhart, Oregon in Clatsop County

*Francis Campbell's ship YFU-62*

where, in 1960, he married Ida Vancampen. Their marriage produced two sons and a daughter.

On January 16, 1969, YFU-62 was pulling away from a concrete ramp and adjacent pier with a supply of 105 millimeter shells when it struck and triggered a 700 pound mine lurking just blow the dark waterline. The explosion broke the back of the boat, its stern soon disappearing onto the river bottom. Eight men died in the blast including Sonny Campbell. On a nearby craft, sailor Mitchel Worsham and an officer witnessed the violent death of YFU-62. "I was looking at it when it happened. It was a horrible thing to witness, but unavoidable. She sank in the stern section in short order, with 2/3rds of her still above water…. About 4hrs. later we started up River again, not reaching the YFU 62 site when a Mike 8 boat hit another mine further up River. It seemed kind of strange that the only time we had our officer on board from … Danang that all this garbage happened. I remember seeing all the color drain from the back of his neck when the 62 went up. This is a recurring memory for me, as it was the baddest day for me in Vietnam." Despite efforts to sweep the river for additional mines, the second freight boat which struck another mine a few hours later killed three more servicemen that day.

The bodies of all eight deceased were recovered from the river. Sonny Campbell's body was returned to Everett where he was interred in Evergreen Cemetery. A child of Sonny Campbell posted a remembrance to a website honoring those who fell in Vietnam:

"My Father. I will forever miss you, unfortunately I never got to know you. I can only hope I have been a good representation of you, and that I have taught my children to love and live a good life, as I expect you would have taught me. You taught me Sacrifice, and value in ideals like Freedom, and family. Mom is now in heaven with you, and I miss you both greatly. Thank you for your sacrifice, and I thank you for my life."

One sailor lamented that it was "a very bad day on the river, but not the last one."

# A Dream or Heaven?
## Jake Laan

A picture of Jake Laan shows a thin young man, with the hint of a smile, below the wispy impression of a mustache. Seated, he leans slightly forward in his green battle fatigues, a swath of light blonde hair sweeping across his forehead, his eyes obscured by a pair of brown circular sunglasses. Undoubtedly buried in an officer's action report are the few details of his death on March 3, 1969 near Binh Duong, South Vietnam. The 21-year-old Specialist was a part of the storied 1st Division, dubbed the Big Red One, as designated on their shoulder patch. He was a member of Company A, 2nd Battalion, 16th Infantry and came into the service as a general vehicle mechanic. The Rangers of the 2nd Battalion were engaged in various missions supporting

*Jacob Laan*

a pacification campaign just north of Saigon in early 1969—the so-called Vietnamization process. All we really know is that on that fateful day Jake was returning from a combat mission when an unspecified explosive device ended his life.

Jacob Clark Laan entered the service in December 1967 and received his training in California and Oklahoma before being shipped to Vietnam on July 14, 1968. His father, Jacob Laan, Sr., was born in Amsterdam, Holland in 1913, immigrating to the U.S. in 1920. He was a butcher by trade having worked in packing houses and shops in Hartford, Everett and Kirkland before marrying Jake's mother Charlotte in 1939. Jake Sr. took over the Home Meat Market in East Stanwood which he ran for several years. Although born in Everett on November 6, 1947, Jake Jr. attended schools in East Stanwood after his family moved here in 1949.

Jake's mother remembered her son as a "happy-go-lucky" person but with a serious side. Classmates recall a slightly wild kid who enjoyed to party while fellow soldiers remember a unique individual who was "well liked by everyone." Jake's mother Charlotte called news of her son's death, the "worst day of our lives." Her faith gave her "miracles" which she later recalled:

"When Jake...died in Vietnam, I prayed to know he was in heaven. The Lord woke me several nights with a vision of a letter from Jake saying, 'for those who believe.' The flowers he had sketched were on it and those words on a scroll that blew away in the wind. Then I was put back to sleep. Another night

126

he awakened me (oh ye of little faith) because I prayed again (the first time wasn't enough). This time the letters a foot high said 'At Onement in Christ' in Capital letters. That's the way I felt when he made himself real to me and I understood perfectly. Look at those words: It means atonement. Then I was put back to sleep." For years afterward the family struggled just to get through those days which followed his death.

No great battle was won the day Jake died—no major operation. It was just a typical sweep through local villages where the unknown could meet your every turn. One of Jake's closest army buddies posted his thoughts and memories of Jake: "Remembering how you loved the Spring, it still hurts to know how you were called away that day in March." He remembered a vivid dream where he was again with Jake. It was a strong dream—in part true and in part fiction, he said. After he awoke, he wondered if it was not a dream, then what? "A glimpse of Heaven?" he asked. "I pray that it be so."

*Military Working Dog in Afghanistan*

# Iraq & Afghanistan
## The War on Terror

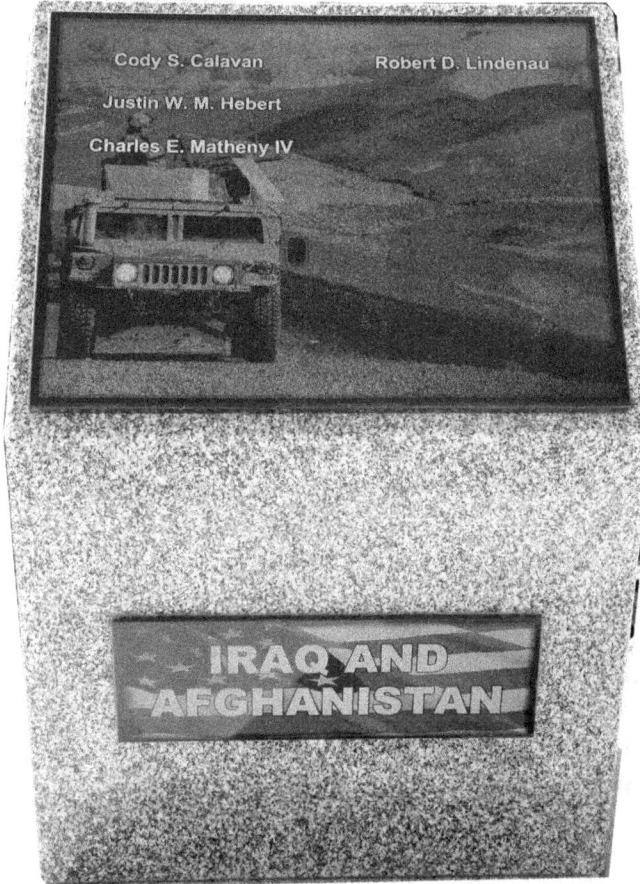

September 11, 2001; a day that would change the future direction of America dramatically. Three commercial aircraft, hijacked by Islamic terrorists, plunged full speed into the twin towers of New York City, the Pentagon in Washington D.C. and a Pennsylvania field after a revolt against their captors by its passengers. Nearly

3,000 people perished in the attacks, the most on American soil since Pearl Harbor. The resulting war on terror would be all-pervasive in the national consciousness for more than two decades after the attacks.

The United States and its allies responded to the invasion of Kuwait by Saddam Hussein with Operations Desert Shield (2 August 1990 – 17 January 1991) and Desert Storm (17 January 1991 – 28 February 1991) which resulted in a defeat for Iraq and liberation of the oil rich Kuwait. Under Operation Desert Shield U.S. forces poured troops into Saudi Arabia as a staging area for the military push into Kuwait. Initial promises to that country to withdraw were tempered by the determination that some forces should remain as a future deterrence to any further threat by Iraq. Within the border of Saudi Arabia lay two of Islam's holiest sites, Mecca and Medina.

Saudi millionaire Osama bin Laden believed that permitting the American infidels to remain was another sign of growing western influence in the Middle East. Bin Laden's religious and political beliefs merged in his obsession to reverse what he saw as greater control of the region by the West. He had supplied money and arms to the Islamic fighters who resisted the Soviet Union's invasion of Afghanistan in the 1980s and maintained his Al-Qaeda organization to continue resistance through symbolic acts of terrorism. Their new embrace of radical jihad included the killing of 18 American servicemen in Mogadishu and the bombing of the World Trade Center in New York City in 1993 and, of course, the deadly attacks of September 11, 2001.

In the pain and aftermath of 911, President George Bush addressed the Congress and the nation on September 20 announcing "Our war on terror begins with Al-Qaeda, but it does not end there. It will not end until every terrorist group of global reach has been found, stopped and defeated." Operation Enduring Freedom was announced on September 25, 2001. On October 7th of that year Britain and the U.S. unleashed air strikes against Al-Qaeda camps in Afghanistan followed on October 19th by the ground war. American allies in NATO announced that they would also send troops in the effort.  On November 13th Kabul fell to the coalition sending Taliban and Al-Qaeda forces into the rural and mountainous areas of that country to continue guerrilla warfare.

Sixteen months later on March 19, 2003 forces of the United States and its coalition invaded Iraq.  They captured Saddam Hussein on December 13, 2003.  After his trial, he was executed December 30th.  Controversy has swirled around the decision to focus on Iraq at that time.  The Bush administration, supported by the British, accused Hussein of stockpiling weapons of mass destruction in violation of international law.  Iraq had been active in this area producing biological and nuclear weaponry from 1962 to 1991 and had used chemical weapons in its war with Iran in the early 1980s.  United Nations' Resolution 1441 demanded the active cooperation of Hussein's government to facilitate inspections seeking information on the alleged illicit weaponry.  The U.S. and Britain accused Iraq of obstructing the inspections and ordered the invasion of the country.  A 2004 Senate report concluded that pre-war statements by the Bush administration were misleading and were not supported by underlying intelligence.  Inspections of suspected areas after the invasion failed to find any evidence of the production or stockpiling of WMDs.

# They Didn't Come Any Better
## Justin Hebert

Justin Hebert had ambitions of seeing the world.  The bonus money he got for joining the army was to go toward college.  His sister Jessica told the *Arlington Times* that they both wanted a way out of the country town where they grew up, although it was who they were—the products of that little town of Silvana.  For Justin the way out was the army, leaving Arlington High School in June 2001 and only after convincing his parents, Bill and Robin, to sign for the 17-year-old.  He was reportedly attracted to the

*Justin Hebert*

military for the money and education benefits. His uncle Dan Hebert said that "he knew he wanted to see the world and go to college," but with the economy at that time "the service is just about the only chance these kids get." He graduated from basic training at Ft. Sill, Oklahoma. He was assigned to Delta Battery, 319th Airborne Field Artillery Regiment of the 173rd Airborne Brigade; the "Sky Soldiers" had been reactivated in 2000. His family said that Justin "thrived in the Army," met a girlfriend in Germany and was earning college credit through internet courses.

Justin William Michael Hebert was born in Everett on July 28, 1983 but knew the little farming community of Silvana as home. He was enthusiastic about sports, especially soccer and baseball, and enjoyed talking snowboarding. Chad Winterhalter, who grew up with Justin, remembered that he "had a knack for making people laugh." That humor was evident in a letter home to Jim Payne, of Willow and Jim's Cafe, where Justin sarcastically warned that, "Iraq is probably the worst place in the world to vacation. I wouldn't recommend it."

Justin completed his training just a week before planes brought down the New York City towers. Initially posted to Vicenza, Italy, he and the 173rd became part of operation Iraqi Freedom in late March 2003 with a parachute drop into northern Iraq, seizing

Bashur Airfield and securing Kirkuk 15 days later. He was among the first combat troops into Iraq. Ironically, and perhaps telling, this war-ravaged area 150 miles north of Baghdad near the Tigris River is considered the "cradle of civilization."

Justin, as a fire-support specialist, operated in forward areas training lasers on targeted enemy sites for Allied aircraft and artillery. Iraqi Freedom was declared a victory on May 1, 2003 but Hebert's unit continued to work enforcing night curfews. Justin was driving a Humvee as the rear vehicle in a convoy near midnight on the evening of August 1st, singing slightly off-key one of his favorite songs, as was his habit. As they maneuvered down the steep decline from the desert ridge toward the bridge, ominously dubbed the "Bridge of Death," to cross over the Tigris River, his unarmored vehicle was peppered with automatic gunfire. Justin punched the accelerator to escape the ambush and save his comrades as a rocket-propelled grenade ripped into the driver's side window. Others in the Humvee were wounded but Justin was killed; the first from Snohomish County to die in Iraq and the 52nd soldier to perish to that point since the government declared the major fighting over.

Once when Justin and Chad visited the Little White Lutheran Church on the Hill outside Silvana, Justin picked out a quiet spot and told his friend that was where he wanted to be buried. On August 16, 2003 an estimated 400 residents gathered to say goodbye as Justin was laid to rest in that spot next to the little church. Sergeant Nicholas Lewis accompanied Justin's body home, fulfilling a promise but also helping salve his own grief. "I ask God to protect him, to keep his music loud and his steaks well-done," he told the Associated Press.

In October 2003 Sergeant Jason Haynes was interviewed as he sat at a patrol site in Iraq near the site where Justin died. He told *Stars and Stripes* that he struggled with the death of Justin Hebert, questioning the war's purpose. Justin, said Haynes, was "the type of guy that always had a smile on his face. And he was a good soldier. They don't come any better."

# Given A Choice
## Cody Calavan

*Cody Calavan*

Nearing the end of his boot camp in 2003, Cody Shea Calavan received word that his younger brother, 15-year-old Joey, had died in an accident near Stanwood involving a drunk driver. As the sole surviving son of his family, Cody could have easily opted for a combat deferment, but letting his fellow Marines deploy to Iraq without him was never a consideration. Confident of himself and his ability he didn't think twice about it, according to his grandfather. Other family members recalled a young man who valued loyalty and service and took great pride in being a Marine even having the Marine dictum Semper fidelis tattooed on his back. He joined the Corps right after graduating from high school in Lake Stevens. His parents moved at that time to Stanwood. After the service, Cody planned to become a police officer.

As part of Company E, 2nd Battalion, 4th Marine Regiment, 1st Marine Division at Camp Pendleton, 19-year-old Pfc. Calavan was deployed as a machine gunner in February 2004 and quickly became involved in the desperate fighting in Al Anbar Province south of Baghdad and particularly the provincial capital of Ramadi. Headlines in the spring of 2004 focused on other hotspots such as Fallujah even as insurgents slipped into Ramadi where in April the 2nd Battalion was subjected to well organized and deadly ambushes. During this time Marines of the 2nd Battalion suffered 34 dead and 255 wounded—a staggering statistic for a six-month deployment. Cody Calavan escaped one such attack thanks to his Kevler vest which absorbed a shot that reportedly knocked him off his feet. The intense April attacks became more sporadic but one reporter wrote that "the Marines of the 2-4 never really had a peaceful day for the rest of their deployment."

Cody became one of the sad statistics of 2-4 on May 29, 2004, ironically part of Memorial Day weekend in the States. Cody and two other Marines died that day when a car bomb mangled their Humvee as it passed along a stretch of Ramadi highway.

Adventurous and mischievous—always operating at full throttle-- is how his sister remembered Cody. "A picture of fitness," recalled friend Jim Eylander who lived near the family in Lake Stevens. Sgt. Ronnie Ramos, who served with Cody, called him a "brilliant Marine" who wanted to be the best. He was one of the leading gunners in the unit and never complained, remembered Ramos. "He had the heart of a great leader," Ramos said. Cody was given a choice said his grandfather and "we're proud of the choice he made." At his memorial service Pastor John Geiszler said that he believed that Cody found purpose in his life and made the most of it. He was a "remarkable young man, said the clergyman, "and left a legacy far beyond his years."

*Washington Post* reporter Thomas Ricks was embedded with troops in Iraq and a veteran military reporter for 15 years. In a June 2004 article he reflected on American losses listing names and facts about those lost up to that time, including Cody S. Calavan.

"He was younger than my own son, I think--born when Ronald Reagan was president, and probably still in kindergarten during the 1991 Persian Gulf War...They are all losses," he wrote, "but the youngest ones haunt me most."

# Even the Sky Wept
## Charles Matheny IV

The strains of two bagpipes mournfully droned "Amazing Grace" as the body of Charles E. Matheny IV was laid to rest on February 27, 2006. The 23-year-old Army sergeant had died only nine days earlier when a roadside bomb shattered his Humvee in Baghdad, Iraq, killing him instantly. Rifle volleys of a 21-gun salute pierced the air as family members and friends wept. The day's drizzle slowly turned to rain prompting one reporter to write that, "even the sky began to cry" as the sound of taps accompanied the folding of the American flag and its presentation to his parents.

The young man listed his hometown as Stanwood although he was born in San Diego on March 12, 1982 and graduated from Arlington High School in 2000, joining the service just one month before the attack on New York's Twin Towers in 2001. Charlie had attended Stanwood Middle School and Stanwood High before finishing school in Arlington where his father lived. His mother and current husband had moved to Camano Island by the time of his death.

Charlie was known for his sense of humor and contagious laugh but it was his talent as a mechanic that got him assigned to Company F, 704th Support Battalion, 4th Infantry Division. His passion for fixing cars included his beloved Mustang GT convertible. He was one of those "part-of-the-solution kind of kids," remembered one of his Arlington teachers.

*Charles Matheny IV*

He was the fourth generation of Mathenys, stretching back to World War I, to serve his country and Charlie dreamed of joining the Army even as a boy. His parents met serving in the same maintenance battalion where their son was later assigned. He was on his second deployment to Iraq when he was killed.

Fox Company supported other units by conducting combat logistics' patrols across central Iraq, delivering water, fuel and spare parts and aid in construction of checkpoint barriers. Charlie's Humvee was the second in a convoy when it rolled past what looked like a rock but instead was an armor-piercing cluster bomb. His father said that the explosion eviscerated the lower half of his son's body, killing him instantly. Chuck Matheny said that Charlie volunteered to assist training Iraqi security forces and was with the Iraqi Army in the dangerous Baghdad slum of Sadr City when he died. His father remembered that Charlie took risks so that others did not have to, replacing fellow soldiers who were married or had children. He was at a stage in life where "he had to rise to the occasion and be the man," his father said. "He was a hero", said his mother Dedi, "My heart is empty, but I'm so very proud of my son."

The pain of Charlie's passing is evident in the newspaper accounts of the event and from the prolific postings on a website in his honor. Some who served with him attended his 2006 funeral before his burial at Tahoma National Cemetery in Kent, Washington. One read a poem written by Charlie's best friend Sgt. Heath Ward:

It seems a dream that you have vanished,

Never to come back again.

For you, Charlie, my thoughts are filled with memories

And my heart is edged in black.


# A Spark of Hope
## Robert Lindenau

Robert Duane Lindenau believed in the human spirit. He thought that where there is hope, all could achieve a better life. As a team leader with the 91st Civil Affairs Battalion, he had the opportunity to put his faith in humanity into action, by working in Afghan medical clinics and helping local families, especially the children of war-torn Afghanistan. Those children—all children—were drawn to him, his uncle Mike Bloom remembered.

Captain Lindenau's mission--a mission he loved--ended on October 20, 2008 when his vehicle was struck by a rocket-propelled grenade in northeastern Afghanistan. He was just two months past his 39th birthday. His promotion to major was made posthumously. He left behind his wife Tonya (formerly Luce/Bloom) and four children. Tonya graduated from Stanwood High School in 1989.

Perhaps it was Bob's musical soul that children responded to. The Camano Island resident was born August 22, 1969 in Seattle but

138

*Robert Lindenau*

attended the University of Idaho where he earned a Bachelor's and Master's degree in music and classical guitar performance in 1992 and 1996. He joined the Army in July 1996 shortly after his graduation. His uncle thought Bob didn't seem the "Army type" but came to appreciate the example that Bob set for the soldiers with whom he served.

Robert rose through the ranks after first being trained as a wheeled-vehicle mechanic. He attended Officer Candidate School in 1999 and his following assignments included company fire support officer and targeting officer for 1st Battalion, 17th Infantry Regiment and an assistant operations officer, 1st Battalion, 37th Field Artillery Regiment. After joining Civil Affairs, he served as a team leader with the 96th Civil Affairs Battalion (Airborne) and lastly as a leader of Civil Affairs with the 91st Battalion. As a Civil Affairs officer—a soldier of peace--his job was to ease some of the suffering the war brought to Afghans such as the farmers in the rural area of CharBagh, Laghman Province; a multi-tribal area 100 miles northeast of Kabul near the border with Pakistan. He served three tours abroad for the U.S. starting in Iraq, then Africa and finally Afghanistan. Among his numerous campaign medals are also a Bronze Star, Purple Heart and Meritorious Service Medal.

His obituary talked of a man "dedicated to his family and faith, which provided him with strength and guidance when he was abroad serving his country. His life was dedicated to the selfless service of others, which enabled him to touch the lives of so many of us. While many people talk, Bob did."

In a eulogy, given by Major Danford Bryant, the officer told the tearful gathering that Bob was one of the nicest people you could ever meet. He "had a deep belief that even in the most desperate times, in the most hostile of locations, in the places of deepest despair, the human spirit only needs a little spark to begin the flame of hope. And with hope, he believed all persons can achieve a better life." He told his own children that he was going to a country to help other children to enjoy the safety and freedoms that U.S. citizens enjoy. He believed the children of Afghanistan needed such hope.

*Robert Lindenau*

Bob asked for two things after his death: a wake with food and good beer and the playing of a favorite piece of music--Bach's Chaconne in d minor. Clearly a man with broad tastes. On the same day that Captain Lindenau died of his wounds, a suicide bomber killed seven others 50 miles away in the Province of Kunduz. Two were German soldiers. The other five were Afghan children.

# Epilogue
## A Posthumous Victory for Veterans' Rights
### Frederick Eglinton

The Stanwood VFW Post 2586 was dedicated in July 1932, coinciding with demonstrations in Washington, D.C. by World War I veterans, dubbed the Bonus Army, demanding early redemption of promised payments as many struggled during the Great Depression. Denied by Congress, President Herbert Hoover ordered their dispersal, accomplished violently by units of the U.S. Army. The VFW and its larger partner, The American Legion (1919) had worked diligently to get the World War Adjusted Compensation Act (the soldier's bonus) passed in 1924.

The Stanwood post was named for a young World War I veteran, Frederick G. Eglinton. When the local post closed in 1982 a small memorial was built just south of Highway 532, in a field adjacent to a Stanwood Park and Ride. Over the years its rusted flag poles and brass nameplate became a nondescript sight to passing motorists. The memorial, however, carried the hopes of its benefactors that people might remember the young man from Bellingham. His father William had lobbied to name a post in the young soldier's honor.

The plaque bore no description of Eglinton's war disability or early death in August 1921 at the age of 23, or the pulmonary tuberculosis which ravaged his body a year before; a disease which Eglinton and his family directly linked to the gas attack in France he endured during World War I. Initially a member of the 2nd Coast Artillery of the Washington National Guard, Eglinton's unit was nationalized as the 65th Coast Artillery shortly after America's entrance into the global conflict. Eglinton was honorably discharged from his unit at Camp Lewis on February 28, 1919 upon his return from Europe. He lived for only another 29 months.

Eglinton filed a claim for insurance benefits shortly after his discharge citing his permanent disability. He had maintained his insurance policy with monthly payments during his military service which was to provide $10,000 in benefits as mandated by the War Risk Insurance Act of 1914 and 1917 and administered by the Treasury Department of the federal government. However, the payments were never honored by the Bureau of War Risk Insurance.

Although too late to benefit Frederick Eglinton, his parents, with the help of the American Legion state Adjutant Jesse W. Drain, sued the government in 1925 for the money owed their son claiming negligence or fraud by the Veterans' Bureau. Drain had been a lieutenant in Eglinton's unit in France and his growing reputation would mark him as one of the leading activists for veterans' rights on the Pacific coast. Through petition William and Trina Eglinton made Drain administrator of their son's estate and, as such, he joined in the lawsuit as a plaintiff. Newspapers claimed the suit the first kind of record in the federal court. The same year as the lawsuit, a columnist for H. L. Mencken's *American Mercury* magazine wrote that, "Congress little realizes that its creature, the Veterans Bureau, has probably made wrecks of more men since the war than the war itself took in dead and maimed."

A year after filing, a federal jury awarded the estate of Frederick Eglinton the full amount of his policy along with a lump sum due the young soldier from the time of his discharge in 1919. William and Trina Eglinton remained active in their efforts on behalf of veterans of northwest Washington. Although Fred Eglinton was not there with them, the lawsuit argued in his name puts him in the ranks of those who fought not only for their country, but for the rights promised them upon their return.

For the thousands of those who have given the ultimate sacrifice, millions return to their homes and families often changed forever by their experiences in defense of this country. Blood rises with the singing of national hymns, the unfurling of flags and the drumbeat of a call to action for the country. The work of caring for the men and women who devoted themselves to the cause often dims in the years following the noise of the immediate crisis. We would do well to mark ourselves as a society not by our fervent response to the heat of a critical American moment but in the compassion we quietly but consistently show as a nation after the shooting has stopped.

*"To care for him who shall have borne the battle and for his widow, and his orphan..."*

*Abraham Lincoln, March 4, 1865*

# Sources

## World War I
### The War to end all Wars

## Introduction
T. Ben Meldrum,  362nd Infantry Association, *A History of the 362nd Infantry.*

## A Village Too Far
Andrew Wik

The Story of the 91st Division, https://www.newrivernotes.com/ topical_history_books_1918_worldwar1_storyof_91st_division. htm; T. Ben Meldrum,  362nd Infantry Association, *A History of the 362nd Infantry*; Files of the Stanwood Area Historical Society Archives; Ancestry.com; William H Mason, Snohomish County in the War…, 1926; Excerpts from  Major General William G. Livesay, *The Story of the Powder River: Let 'Er Buck*, 91st Infantry Division, August, 1917-January, 1945,  posted by Chris Mulholland, https:// www.worldwar1centennial.org/index.php/commemorate/ family-ties/stories-of-service/2119-harry-bernard-mulholland. html; Timothy Brown, "The Charge into Gesnes: Sept. 29, 1918," posted  September 2018, http://fieldsoffriendlystrife. com/2018/09/29/the-charge-into-gesne-september-29-1918/; Duane Colt Denfeld, PhD, Wild West Division: "Washington in World War I," Historylink Essay 10648, https://www.historylink. org/File/10648.

## Two Boys from Norman:
Bert Stevens and Andy Engebretsen

*Oregonian*, April 21, 1919; *Harper's Pictorial Library of the World War*, v. 5, Albert Bushnell Hart, (ed.), 1920; William H Mason, Snohomish County in the War…, 1926; Bryan L. Woodcock, Major, U.S. Army, *The 91st Infantry In World War I–Analysis Of An AEF Division's Efforts To Achieve Battlefield Success*, http:// militarymuseum.org/91stDivWWI.pdf;  Harold H. Burton, *600 Days' Service: A History of the 361st Infantry Regiment of The United*

*States Army, 1919; Illustrated History of Skagit and Snohomish Counties...,* 1906; George Seldes, *You Can't Print That!: The truth behind the news, 1918-1928,* 1929; Captain Roger Heller, *The 361st Infantry Regiment, 1917-1955,* 1955; Files of the Stanwood Area Historical Society Archives; Ancestry.com.

## Stanwood's Symbolic Soldier
Frank Hancock

American Battle Monuments Commission, 28th Division: Summary Operations in the World War, 1944, https://www. worldwar1centennial.org/images/Pennsylvania/28th_Division_ Summary_of_Operations_in_the_world_war.pdf; The Association of the 110th Infantry, History of the 110th Infantry (10th Pa) of the 28th Division U.S.A., 1917-1919, 1920; Ancestry.com; Files of the Stanwood Area Historical Society Archives; *Twin City News,* September 24, 1944.

## Willing to Pay the Price
Joseph Bruseth

Thomas Fleming, Meuse Argonne Offensive of World War I, https://www.historynet.com/meuse-argonne-offensive-of-world-war-i.htm, first published in *Military History* magazine, October 1993; Fleming, The Two Argonnes, https://www.historynet. com/the-two-argonnes.htm, first published in The *Quarterly Journal of Military History,* spring 2018; Letter Albert P. Schad, October 8, 1919, letter Colonel Charles C. Pierce, March 11, 1920 to Eliza Bruseth, courtesy of the Ericson family; William H Mason, *Snohomish County in the War...,* 1926; *The Reader's Companion to Military History,* 1996, Robert Cowley & Geoffrey Parker (eds.); Files of the Stanwood Area Historical Society Archives; Ancestry. com; Harold Pierce quoted in Edward Lengel, *To Conquer Hell: The Meuse-Argonne, 1918.*

## Argonne Hellscape
Alfred Kristoferson

Beth Fortson, Spotlight: The Pilgrimages of Gold Star Mothers and Widows, September 27, 2016, https://unwritten-record.blogs.

archives.gov/2016/09/27/spotlight-the-pilgrimages-of-gold-star-mothers-and-widows/; Emil B. Gansser, *History of the 126th Infantry in the War with Germany,* 1920; Thomas Fleming, Meuse Argonne Offensive of World War I, https://www.historynet.com/meuse-argonne-offensive-of-world-war-i.htm, first published in *Military History* magazine, October 1993; The Côte De Châtillon, https://www.pbs.org/wgbh/americanexperience/features/macarthur-ww1-cote-de-chatillon/; United States Department of the Interior National Park Service National Register of Historic Places Registration Form, https://dahp.wa.gov/sites/default/files/NR_nom_KristofersonDairy.pdf; Charles Rhuel Myers, Captain 126th Infantry, Story of the 126 Infantry Regiment, https://www.worldwar1centennial.org/index.php/michigan-in-ww1-articles/1796-the-story-of-the-126th-infantry-regiment.html; Richard S. Faulkner, Meuse-Argonne: 26 September-11 November 1918 U.S. Army Center of Military History, https://history.army.mil/catalog/pubs/77/77-8.html; *Seattle Daily Times,* April 2, 1931; Files of the Stanwood Area Historical Society Archives; Ancestry.com.

## An Immigrant for America
Jacob Teiseth

*Stanwood Tidings,* December 27, 1918; *The Spokesman Review* (Spokane, WA) December 27, 1918; *Seattle Daily Times,* August 26, 1919; A. E. Crane, The 6th Engineers in the Meuse-Argonne. *The Military Engineer,* v. 23 no. 128, March-April 1931; *History of the 6th Engineers and Its Men,* 1920; The United States World War I Centennial Commission, Diseases in World War I: Infectious Diseases, https://www.worldwar1centennial.org/index.php/diseases-in-world-war-i.html; Arlene Balkansky, The Draft in World War I: America "Volunteered its Mass," https://blogs.loc.gov/headlinesandheroes/2018/06/wwi-draft/, June 19, 2018.

## Laid in a Soldier's grave.
Albert Buli and Ray Bunton

Professor John W Graham, *The Gold Star Mother Pilgrimages of the 1930s: Overseas Grave Visitations,* 1961; Richard S. Faulkner, Meuse-Argonne: 26 September-11 November 1918 U.S. Army Center of

Military History, https://history.army.mil/catalog/pubs/77/77-8.
html; The United States World War I Centennial Commission, A
Brief History of U.S. Marine Corps Action in Europe During World
War I, https://www.worldwar1centennial.org/index.php/usmc-
in-ww1/850-a-brief-history-of-u-s-marine-corps-action-in-europe-
during-world-war-i.html; Major Edwin N. McClellan, USMC,
*The United States Marine Corps in the World War,* Marine Corps
Historical Branch, 1920, https://www.marines.mil/Portals/1/
Publications/The%20United%20States%20Marine%20Corps%20
in%20the%20World%20War%20%20PCN%2019000411300.pdf;
findagrave, https://www.findagrave.com/memorial/18841517;
*Aberdeen Daily News,* November 5, 1918; *Seattle Star,* October 26,
1918; *Lynden Tribune,* November 28, 1918; Michael A. Eggleston,
*The 5th Marine Regiment Devil Dogs in World War I,* 2016; Files of
the Stanwood Area Historical Society Archives; Ancestry.com.

## As If God Was Calling Them
Emma Thorsen

*Evening times-Republican,* February 18, May 30, June 18, Sept.
30, Oct. 19, 1918; *Ottumwa Semi-Weekly Courier,* Jan. 15, 1918;
*Denison Review,* April 10, April 24, Oct. 30, Aug. 21, 1918;
*Bottineau* [N.D.] *Courant,* Oct. 10, 1918; Robb letter in Frank
Santiago, *Des Moines Register,* November 1999, digital re-release
March 11, 2020, https://www.desmoinesregister.com/story/
news/local/2020/03/03/influenza-outbreak-700-soldiers-died-
1918-iowa-camp-dodge-johnston-army-base-national-guard-
flu/2843939001/; *Des Moines Register,* August 19, 1983; *Journal of
Minnesota Medicine,* v. II. No. 8, August 1919; Carol R. Byerly, "The
U.S. military and the influenza pandemic of 1918-1919." Public
health reports (Washington, D.C. : 1974) vol. 125 Suppl 3, Suppl
3 (2010): 82-91; Dan Vergano, "1918 Flu Pandemic That Killed 50
Million Originated in China, Historians Say," *National Geographic
Magazine,* digital publication January 24, 2014; G. Dennis Shanks,
et al, Low but Highly Variable Mortality Among Nurses and
Physicians During the Influenza Pandemic of 1918-1919, https://
www.ncbi.nlm.nih.gov/pmc/articles/PMC4941589/; Ancestry.
com; Files of the Stanwood Area Historical Society Archives;
Ancestry.com.

# World War II
## The War to Defeat Fascism

## Introduction

*Twin City News*, January 7, 1945, April 26, 1945; Scrapbook, n.d., Files of the Stanwood Area Historical Society; Alice Essex, *Stanwood Story*, V. 3.

## A Seeker of Wings
### Alvin Vaara

*Dallas Morning News*, January 25, 1943; John Young, *Waco Tribune-Herald*, digital copy, http://wacohistoryproject.org/Places/airfields.htm; Ancestry.com.

## A Soldier's Death
### Robert Baker

"Rolling Ahead!" a small booklet covering the history of the 89th Infantry Division. This booklet is one of the series of G.I. Stories published by the Stars & Stripes in Paris in 1944-1945.-- http://www.lonesentry.com/gi_stories_booklets/89thinfantry/; Captain Joe D. Hennessee, The Operations of the 2D Battalion, 353rd Infantry (89th Division) in the Assault Crossing of the Moselle River in the Vicinity of Bullay, Germany, and Subsequent Pursuit Action, 16 – 20 march, 1945, https://mcoepublic.blob.core.usgovcloudapi.net/library/DonovanPapers/wwii/STUP2/G-L/HennesseeJoeD%20CPT.pdf; Ancestry.com.

## Semper Fidelis
### Charles Isham

http://wc.rootsweb.ancestry.com/cgi-bin/igm.cgi?op=GET&db=ishamsplit&id=I4352; http://focus.nps.gov/pdfhost/docs/NRHP/Text/01000505.pdf; http://dezdan.com/washington/paradise-river-hydroelectric-plant/; *San Diego Evening Tribune*, March 25, 1932, February 26, 1935, November 9, 1938; *Ellensburg Daily Record*, June 5, 1943; *Seattle Daily Times*, May 25, 1942; *Twin City News*, January 15, 1948; Lt. Edward F. O'Day, Report of Engagement with Bandit Forces, Condega (El Bramadero, 27

Feb. 1928), posted by Michael J. Schroeder, PhD, http://www.
sandinorebellion.com/PCDocs/1928a/PC280301-O'Day.html;
Files of the Stanwood Area Historical Society Archives; Ancestry.
com.

## Onward Christian Soldier:
### The Final Mission of Gerhard Lane

*Seattle Daily Times*, June 6, 1935, May 21, 1938, April 1, 1943, March
5, 1945, August 12, 1946 and August 20, 1949; *Bellingham Herald,*
May 21, 1924, May 20, 1935, April 1, 1943; *Tacoma Daily Ledger,*
May 22, 1927; *Tacoma News Tribune*, May 14, 1927, May 22, 1929;
*Minneapolis Star*, April 13, 1943; Edward Kuder and Pete Martin,
The Philippines Never Surrendered, *Saturday Evening Post*, Feb. 10,
17 and 24, 1945; "Experience of a Supply Officer," excerpts from
report of Commander Walter Bicknell, USN, oryokumaruonline.
org; Pacific Lutheran University Archives and Special Collections;
The Esaches, Stanwood High School yearbook, 1925, 1926,
Stanwood Area Historical Society Archives; http://www.
usmilitariaforum.com/forums/, accessed March 3, 2020; Ancestry.
com.

## A Short but Devout Life
### Roger William Barney

*Twin City News* from the files of the Stanwood Area Historical
Society Archives; *Bellingham Herald*, June 6, 1944; Ancestry.com.

## "Death Was No Stranger"
### Harold McCann

Prisoners of War in the Philippine Islands Military Intelligence
Division Report September 20, 1944, online; Commander
Melvyn McCoy, et al, "Death Was Part of Our Life," How 5200
Americans and Thousands of Filipinos Died in Jap Prison Camps,
*Life Magazine*, February 7, 1944; Military Intelligence Division
Report, September 20, 1944, Prisoners of War in the Philippine
Islands, http://www.mansell.com/pow_resources/camplists/
philippines/pows_in_pi_report.html; Files of the Stanwood Area
Historical Society Archives; Ancestry.com.

## Tablets of the Missing
Donald Garrison

Account of Captain Daniel T. Roberts, https://pacificwrecks.com/aircraft/p-38/42-66547.html; Account of Lieutenant Thomas J. Simms, https://pacificwrecks.com/aircraft/p-38/42-66833.html; Wikipedia; Files of the Stanwood Area Historical Society Archives; 475th Historical Foundation, https://475th.org/history/unithistory/; Ancestry.com; Garrison family letters and clippings, courtesy of Donald Garrison's son, Donald Garrison Buss.

## A Self Made Man
Leonard Broin

*Tatooed on My Soul: Texas Veterans Remember World War II,* Stephen Sloan, et al, (eds.), Texas A & M Press, 2015; Auxiliary Motor Minesweepers (YMS), http://www.navsource.org/archives/11/19idx.htm.

## Death Along the Rapido
Peter Rekdal

Bethanne Kelly Patrick, Rapido River Disaster, https://www.military.com/history/rapido-river-disaster.html; *Twin City News,* May 1943, July 8, 1943, March 30, 1944; Files of the Stanwood Area Historical Society Archives; Ancestry.com.

## Hell Ship Horror
Ernest Moser

American Prisoners of War in the Philippines, Office of the Provost Marshal General Report, November 19, 1945, http://www.mansell.com/pow_resources/camplists/philippines/pows_in_pi-OPMG_report.html; Lee A. Gladwin, American POWs on Japanese Ships Take a Voyage into Hell, *Prologue Magazine,* Winter 2003, Vol. 35, No. 4, http://www.archives.gov/publications/prologue/2003/winter/hell-ships-1.html; Ancestry.com; *Twin City News,* September 9, 1943; Ken Wheeler, Activity: We Did Not Surrender: The POW Experience in the Philippines, https://www.

nhd.org/sites/default/files/We%20Did%20Not%20Surrender%20 -%20Lesson%20Plan.pdf; C. Peter Chen, Shinyo Maru, https:// ww2db.com/ship_spec.php?ship_id=530; Files of the Stanwood Area Historical Society Archives; Ancestry.com.

## The Deadly Blossoms
### Gordon Lord

American Air Museum, http://www.americanairmuseum.com/ aircraft/15042; Stewart Halsey Ross, *Strategic Bombing by the United States in World War II: the Myths and the Facts*, 2003; Files of the Stanwood Area Historical Society Archives; Ancestry.com.

## Extraordinary Valor
### Robert Nelson

Eric M. Hammel, *Air War Pacific Chronology : America's Air War against Japan in East Asia and the Pacific, 1941-1945*, 1998; Edward Peary Stafford, *The Big E: The Story of the U.S.S. Enterprise*, 2002; *Bellingham Herald*, April 7, 1924, October 14, 1929, March 10, 1944; Ancestry.com including Battle family genealogy; http://www. worldwarphotos.info/gallery/usa.aircrafts-2-3/f6f-hellcat/ens-bob-nelson-vf-20-ace-by-f6f-5-hellcat/.

## A Cottage at Warm Beach
### Donald Leach

*Arlington Times*, December 12, 1935; *Seattle Daily Times*, July 5, 1946; U.S. Army, 2nd Infantry, Second Infantry Regiment, Fifth Infantry Division, 1946, https://books.google.com/ books/about/Second_Infantry_Regiment_Fifth_Infantry. html?id=lxfpAAAAMAAJ; Files of the Stanwood Area Historical Society Archives; Ancestry.com.

## Two Gold Stars
### Dorman Riker and Andrew Riker

William Bowen, THE ARISAN MARU TRAGEDY, www,us-japandialogueonpows.org; The *Arlington Times*, April 27, 1933, July 8, 1933, April 13, 1939, October 17, 1940; Karen Ann Takizawa, War Stories: Bilibid Prison and the Hellships, July 1942 – Jan.

1945, *Hosei Journal of Sociology and Social Sciences,* v. 60, no 3, 2012-2013, posted 2013; Agnes Molstad, *The Victoria Community*, 1976; Ancestry.com; Images courtesy of Mary Jo Porter.

## Held in High Esteem
Edward Pearson

Tom Swope interview of James Anderson, March 12, 2004, http://memory.loc.gov/diglib/vhp/story/loc.natlib.afc2001001.20422/transcript?ID=sr0001; pubmed.ncbi.nlm.nih.gov; Files of the Stanwood Area Historical Society Archives; Ancestry.com.

## "First In, Last Out"
Floyd Perin

65th Engineer Battalion, https://www.globalsecurity.org/military/agency/army/65eng.htm;Wikipedia; www.warfarehistorynetwork.com; Britannica.com; The 25th Division and World War II, Capt. Robert F. Karolevitz (ed)., 2012; *Bellingham Herald*, July 7, 1925; Files of the Stanwood Area Historical Society Archives; Ancestry.com.

## Home Front Tragedy
Orville Knutson

Stanwood High School yearbook, 1943; Files of the Stanwood Area Historical Society Archives; Ancestry.com; Wikipedia.

## War Darkens a Local Home
Vivien Mickel

*Seattle Daily Times*, July 17, 1943, April 23, 1944, January 5, 1945, May 2, 1947, May 11, 1949, *Twin City News*, February 8, March 15, April 26, November 29, 1945; William H. Boudreau, 12th Cavalry History, https://1cda.org/history/history-12cav/; Files of the Stanwood Area Historical Society Archives; Ancestry.com.

## "Pushing Forward"
Daniel Hess

Eugene Sledge, *With the Old Breed: At Peleliu and Okinawa*, 1981; Capt. James R. Stockman USMC, *The First Marine Division on*

*Okinawa; 1 April - 30 June 1945;* Files of the Stanwood Area Historical Society Archives; Ancestry.com.

## "Victims of Perdition"
Edwin Lund

Richard Sams, Perdition: A Forgotten Tokyo Firebombing Raid, *Asia Pacific Journal,* v. 14, Issue 12, June 15, 2016, https://apjjf.org/2016/12/Sams.html; Anthony Pomata, Boeing B-29 Superfortress Bomber, HistoryLink.org Essay 3828, 6/05/2002; Polly Reed Myers, Boeing Aircraft Company's Manpowr Campaign During World War II, *Pacific Northwest Quarterly,* v. 98, no. 4, Fall 2007, jstor.org; http://sites.rootsweb.com/~ny330bg/crews458.htm.

## "A Damn Long Day"
Arne Aalbu

James A. Huston, *Biography of a Battalion: The Life and times of an Infantry Battalion in Europe in World War II,* 2011; https://vetaffairs.sd.gov/sdwwiimemorial/SubPages/profiles/Display.asp?P=1; Captain Raymond J. Anderson, 134th Morning Report for December 8, 1944, http://www.coulthart.com/134/mr-l-company/mr-134-l-1944-12-8.pdf; Anderson, 134th Morning Report for December 9, 1944, http://www.coulthart.com/134/mr-l-company/mr-134-l-1944-12-9.pdf; http://www.coulthart.com/134/uj/uj-134-1944-december.pdf.; www.coulthart.com/134/134-ir/aalbu-arne-o.htm.

## Death by Divine Wind
Robert Harrison

Robert Stern, *Fire From The Sky: Surviving the Kamikaze Threat,* 2010; Wayne Haviland, Account of the Sinking of the USS REID DD369 - December 11, 1944, http://ussreid369.org/warstories.htm; Wikipedia, USS Reid (DD-369); Ancestry.com.

## A Deadly Christmas
Robert Bransmo

Donald Shaub, interview by Joe Todd, Veteran recalls Leopoldville struck by torpedo, Part 1, in *Bartlesville* (OK) *Examiner Enterprise* June 2, 2009, posted April 2, 2012, https://www.examiner-enterprise.com/news/local-news/bartians-recall-wartime-experiences-anticipate-honor-flight; SS Leopoldville Wreck, https://divescover.com/dive-site/ss-leopoldville-wreck/30928, Ancestry.com.

## The Last Shot
Wesley Sigerstad

History of the 97th division during WWII, http://www.97thdivision.com/historyp1.html; "Last Shot" Memorial, Fort Benning, Georgia, http://www.97thdivision.com/lastshot.html; Files of the Stanwood Area Historical Society Archives; Ancestry.com.

## The Peace Medal
Richard Hiday

The *Oregon Statesman*, February 11, 1943, August 27, 1943, February 25, 1944, January 6, 1946, December 22, 1948, January 28, 1949; *Daily Capital Journal* (Salem, OR), August 7, 1945, January 7, 1946; Ancestry.com.

# Korea
## Containment of Communism

## Introduction
*Twin City News*, September 14, 1950

## Pure Survival
Harold Horton

Charles J. Hanley, Martha Mendoza and Sang-Hun Choe, GIs: U.S. killed Korean War refugees, Report details more instances of civilian deaths at hands of U.S. in Korean War, Associated Press, posted October 14, 1999; Files of the Stanwood Area Historical Society Archives; Ancestry.com.

## They Died Unknown
Earl Christensen

Jack D. Walker, A Brief Account of the Korean War, Korean War Veterans Association, Inc., http://www.kwva.org/brief_account_of_the_korean_war.htm; Encyclopedia of the Korean War: A Political, Social, and Military History, Volumes 1-3, Spencer C. Tucker, Paul G. Pierpaoli Jr. (eds.), 2000 ; Files of the Stanwood Area Historical Society Archives; Ancestry.com; www.63rdinfdiv.com/miscphotospage76.html.

## "Don't Worry About Me, Mom"
Clarence Hartley

Edward L. Daily, *Skirmish Red, White and Blue, The History of the 7th U.S. Cavalry, 1945-1953*, 1992; Lt. Col. Roy E. Appleman, *Disaster in Korea: The Chinese Confront MacArthur,* 1989; Strike Swiftly Korea, 1950-1953: 70th Heavy Tank Battalion, Turner Publishing, 1988; *1st Cavalry Division: A Spur Ride Through the 20th Century From Horses to the Digital Battlefield*, Herbert C. Banks (ed.), 2003; Files of the Stanwood Area Historical Society Archives; Ancestry.com.

# Vietnam
## Countering the Dominoes

## Outside the Wire
Warren Gary Peterson

History of the 327th Infantry Regiment, https://www.327infantry.org/bastogne/327th-infantry-regiment-history/history-of-the-327th-infantry-regiment/; Postings by Steve Good and Dennis Wriston, https://www.vvmf.org/Wall-of-Faces/40494/WARREN-G-PETERSON/; David H. Hackworth, *About Face: The Odyssey of an American Warrior*, 1990; Michael Sallah and Mitch Weiss, *Tiger Force: A True Story of Men and War*, 2006; *Evening Star*

(D.C.), February 10, 1966; *Aberdeen Daily News* (S.D.), February 10, 1966; ; Files of the Stanwood Area Historical Society Archives; Ancestry.com.

## A Sky Soldier Falls
Frank Smith

Citation for First Lieutenant Daniel J. Severson, https://valor. militarytimes.com/hero/5516; findagrave; Frank Lee Smith, 173rd Airborne Brigade, http://www.alch372.com/503/frankleesmith. htm; News article posted on https://careersdocbox.com/US_ Military/77199852-2-503d-photo-of-the-month.html; http://www. honorstates.org/index.php?id=298764; Files of the Stanwood Area Historical Society Archives; Ancestry.com.

## "A Day of Extreme Sacrifice"
George Broz

Michael Thill, Recollections from late Christmas 1967: A day of extreme sacrifice in the Vietnam history of 3/1, http:// militarysignatures.com/signatures/member1520.png; Do not stand at my grave and weep, posted by Bob Ross, December 27, 2008, https://www.vvmf.org/Wall-of-faces/6350/GEORGE-M-BROZ/#sthash.0BXvSYaB.dpuf; Bradford A. Wineman, Ambush at Thon Tham Khe, https://www.historynet.com/ambush-thon-tham-khe.htm; Letter George Broz to Elizabeth Broz, December 9, 1967, provided to the author by the Broz family; Recollections of Lee Ayers, Stanwood eighth grade teacher on George Broz, e-mail communication with Jim Joyce, April 17, 2015; Files of the Stanwood Area Historical Society Archives; Ancestry.com.

## Beyond the Call of Duty
Elliott Peters

Paul "Hungry" Marquis, Anatomy of a Fire Mission, http://www. willpete.com/Firebases/anatomy_of_a_fire_mission.htm; Charles R. Smith, *A Brief History of the 12th Marines*, 2014; Letter from Jim Peters (brother of Elliot) to his sisters, provided to the author via e-mail by his sister Leslie Asbury, December 24, 2015; After action report of Battery B, 1st Battalion, 12th Marines; https://

www.businessinsider.com/how-dale-dye-earned-bronze-star-in-vietnam-2015-3; http://www.vvmf.org/Wall-of-Faces/40383/ELLIOTT-L-PETERS/page/2/; Files of the Stanwood Area Historical Society Archives; Ancestry.com.

## "A Very Bad Day On The River"
Francis Campbell

James Steffes, *Swift Boat Down: The Real Story of the Sinking of PCF-19,* 2005; Mike Rice, http://www.yosemite-photo.com/guestbook/YFUgb.htm; Mitchel Worsham, www.military.com/homepage.unitpagememberprofile1,13480; www.navsource.org/archives/10/18/1018097304.jpg; Files of the Stanwood Area Historical Society Archives; Ancestry.com.

## A Dream or Heaven?
Jake Laan

Jake Clark Laan, https://www.honorstates.org/index.php?id=283141; Alan Duyff, https://www.virtualwall.org/dl/LaanJC01a.htm; Charlotte Clark Laan, https://www.wikitree.com/wiki/Clark-26412; "Spirit in the Sky" by Bob Hodge, November 12, 2018, http://richardhowe.com/2018/11/12/spirit-in-the-sky-by-bob-hodge/; History—the society of the 1st Infantry Division, https://www.1stid.org/history; http://www.vvmf.org./Wall-of-faces/29270/JACOB-C-LAAN/; Files of the Stanwood Area Historical Society Archives; Ancestry.com.

# Iraq and Afghanistan
## The War on Terror

## They Didn't Come Any Better
Justin Hebert

Kelly Ruhoff, *Stanwood Camano News,* August 5, 2003; Chris Maag, Town mourns intent young man killed in Iraq, *Seattle Times,* August 17, 2003; Tony Dondero, *Arlington Times,* August 11, 2004; Kirk Boxleitner, *Arlington Times,* October 18, 2006; Jon R. Anderson, In Iraq, fear is a constant companion, *Stars And Stripes,*

October 19, 2003, https://www.stripes.com/news/in-iraq-fear-is-a-constant-companion-1.12754; Robert Burns, In Iraq, youngest U.S. troops bore heaviest toll, Associated Press, August 23, 2011, https://www.mprnews.org/story/2011/08/20/iraq-war-deaths; Associated Press posted August 20, 2011; https://thefallen.militarytimes.com/army-spc-justin-w-hebert/256769.

## Given A Choice
Cody Calavan

Thomas E. Ricks, *Stamford* (CT) *Daily Advocate,* June 15, 2004; David Hasemyer, *San Diego Union-Tribune,* Thursday, June 10, 2004; *Everett Daily Herald,* June 10, 2004; Kelly Ruhoff, *Stanwood Camano News,* June 22, 2004; Christopher Schwarzen and Peyton Whitely, *Seattle Times,* June 2, 2004; https://thefallen.militarytimes.com/marine-pfc-cody-s-calavan/257296; Katherine Schiffner, A Marine to the end, *Everett Herald,* https://www.heraldnet.com/news/a-marine-to-the-end/; www.findagrave.com/memorial/893776/cody-shea-calavan; Files of the Stanwood Area Historical Society Archives.

## Even the Sky Wept
Charles Matheny IV

*Stanwood Camano News,* February 28, 2006; *Everett Daily Herald,* February 23, 2006, February 28, 2006; Scott Morris, *Everett Daily Herald,* February 27, 2006; *Seattle Times,* February 25, 2006; http://thefallen.militarytimes.com/army-sgt-charles-e-metheny-iv/1550712; Files of the Stanwood Area Historical Society Archives.

## A Spark of Hope
Robert Lindenau

Civil Affairs remembers a friend, https://www.shadowspear.com/vb/threads/civil-affairs-remembers-a-friend.3075/; https://www.fallenheroesproject.org/united-states/robert-duane-lindenau/; https://thefallen.militarytimes.com/army-maj-robert-d-lindenau/3783330; Fr. Phillip Bloom, Funeral Homily for Major Robert D. Lindenau, November 1, 2008, http://

stmaryvalleybloom.org/homilyforboblindenau.html; Gale Fiege, *Everett Daily Herald*, October 20, 2008, https://www.heraldnet. com/news/soldier-killed-in-afghanistan-loved-his-family-music; http://oldcamano.net/captainbob.html; Files of the Stanwood Area Historical Society Archives.

# Epilogue
## A Posthumous Victory for Veterans' Rights
## Frederick Eglinton

*The Bellingham Herald*, 1921-1945; Federal census records, 1900-1920; Alice Essex, *The Stanwood Story*; Web posts from the Whatcom County Genealogical Society.

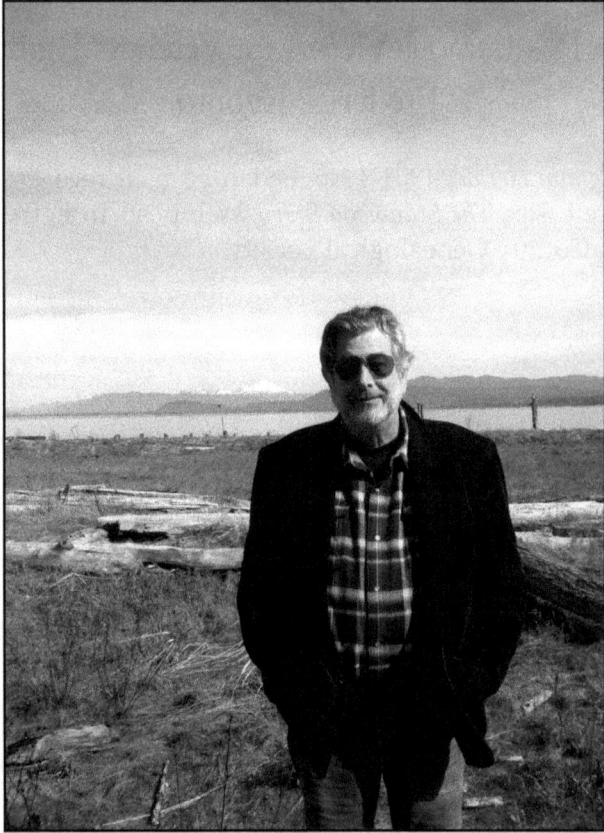

*Richard Hanks at the English Boom Historical Park*
*on the north end of Camano Island, Washington*

# About the Author

Born and raised in the Midwest, Richard Hanks came of age in Southern California where he worked as a journalist, curator, archivist, professor, editor and author. His academic fields of interest include 19th and 20th century American history with particular focus on Native American history for which he earned his PhD at the University of California, Riverside. Upon retiring in 2013 he moved to Camano Island, Washington where he volunteered at the Hibulb Cultural Center on the Tulalip Reservation and as a writer and researcher for the Stanwood Area Historical Society. He was president of the Society from January 2017 to December 2020. He and his wife Robin also began Coyote Hill Press in 2013. His other books are *The Harris Company* (co-author), Arcadia Press, 2008; *This War is for a Whole Life: The Culture of Resistance Among Southern California Indians, 1850-1866*, Ushkana Press, 2012; *Vermont's Proper Son: The Letters of Soldier and Scholar, Edwin Hall Higley, 1861-1871*, Coyote Hill Press, 2014; *A Living History of the Soboba Band of Luiseño Indians and their Connections to the San Jacinto Valley* (co-author), Soboba Cultural Resource Department, 2018.

COYOTE HILL PRESS

www.ingramcontent.com/pod-product-compliance
Lightning Source LLC
Chambersburg PA
CBHW072247270326
41930CB00010B/2294